T0191834

Intelligent Systems, Control and Automation: Science and Engineering

Volume 99

Intelligent Systems, Control and Automation: Science and Engineering book series publishes books on scientific, engineering, and technological developments in this interesting field that borders on so many disciplines and has so many practical applications: human-like biomechanics, industrial robotics, mobile robotics, service and social robotics, humanoid robotics, mechatronics, intelligent control, industrial process control, power systems control, industrial and office automation, unmanned aviation systems, teleoperation systems, energy systems, transportation systems, driverless cars, human-robot interaction, computer and control engineering, but also computational intelligence, neural networks, fuzzy systems, genetic algorithms, neurofuzzy systems and control, nonlinear dynamics and control, and of course adaptive, complex and self-organizing systems. This wide range of topics, approaches, perspectives and applications is reflected in a large readership of researchers and practitioners in various fields, as well as graduate students who want to learn more on a given subject.

The series has received an enthusiastic acceptance by the scientific and engineering community, and is continuously receiving an increasing number of high-quality proposals from both academia and industry. The current Series Editor is Kimon Valavanis, University of Denver, Colorado, USA. He is assisted by an Editorial Advisory Board who help to select the most interesting and cutting edge manuscripts for the series:

Springer and Professor Valavanis welcome book ideas from authors. Potential authors who wish to submit a book proposal should contact Thomas Ditzinger (thomas.ditzinger@springer.com)

Indexed by SCOPUS, zbMATH, SCImago.

More information about this series at http://www.springer.com/series/6259

Shunji Manabe · Young Chol Kim

Coefficient Diagram Method for Control System Design

Shunji Manabe
Fujisawa, Japan

Young Chol Kim
Department of Electronic Engineering
Chungbuk National University
Cheongju, Korea (Republic of)

ISSN 2213-8986 ISSN 2213-8994 (electronic)
Intelligent Systems, Control and Automation: Science and Engineering
ISBN 978-981-16-0548-2 ISBN 978-981-16-0546-8 (eBook)
https://doi.org/10.1007/978-981-16-0546-8

This Springer imprint is published by the registered company Springer Nature Singapore Pte Ltd.
The registered company address is: 152 Beach Road, #21-01/04 Gateway East, Singapore 189721, Singapore

Preface

This book is written with two objectives. The first objective is to introduce a simple but powerful control design technique called Coefficient Diagram Method (CDM), whereby ordinary engineers without strong control and mathematics background can design a good controller for their specific plants. Also, control experts can solve such complicated design problems, which defy their best knowledge, in a consistent manner. The second objective is to clarify the "meaning" of the control theories currently in use. CDM is so general that any controller designed by various control theories can be expressed by an equivalent CDM design. Such equivalent design helps to clarify the "meaning" of each control theory by the common terminology used in CDM. As the result, control experts can correctly evaluate the merits and shortcomings of such control theories and use them wisely in practical application.

Various control theories have been developed so far, and many books are already written. However, the author feels that they hardly come up to the expectation of control designers. Firstly, a strong mathematical background is necessary to understand control theories. The emphasis is placed on analysis, and the design/synthesis problem is not properly addressed. Too much emphasis is placed on Linear Time Invariant (LTI) system, and the extension to time variant and nonlinear systems is not properly considered. For these reasons, the controller designed by such theories is not necessarily a good controller in a practical sense. Secondly, control theories have developed as the answers to specific needs of the industry more or less in an ad hoc manner under the strong influence of the computational power available at that time. Thus, they are so much diversified and lack unity and consistency. The standard textbooks describe these theories as they are in an exhaustive manner. Thus, it is very difficult to understand the meanings of these theories and to grasp the total picture of control theories.

This book is motivated by the need to remedy these situations. Although CDM emerged as a new control design technology in 1991, the basic ideas have been proposed and tested in practical application since the 1950s. The CDM only gives convenient expression and a mathematical basis to such practical ideas. The Kessler standard form [1] has been widely used in the steel industry since 1960 as the model of a good characteristic polynomial. The CDM adopts the Kessler standard form after some modification. The sufficient condition for stability by Lipatov [2], the simplification of the Routh stability condition, is not well known in the control community. However, the theory by Lipatov is very simple but effective and gives the design standard of the characteristic polynomials. The theory constitutes the mathematical basis of CDM. The center of the CDM is the coefficient diagram. The coefficient diagram shows the coefficients of the characteristic polynomial in logarithmic scale (ordinate) and the order of each coefficient in linear scale (abscissa). Its curvature represents stability. Its inclination represents response speed. The deformation of the diagram corresponding to the variation of a specific parameter shows robustness. Thus, the three key elements of the control system, namely stability, response, and robustness, are shown in a single diagram. Because the characteristic polynomial is the sum of the denominator and numerator polynomials of the open-loop transfer function, such polynomials are also shown in the coefficient diagram. With the help of these polynomials, the frequency response and time response can be roughly estimated. Thus, the coefficient diagram contains rich information about the system in a single diagram with an intuitively understandable visual form. In a sense, the coefficient diagram plays the same role as the Bode diagram in a more effective way for design/synthesis problems. The CDM belongs to an algebraic approach, which is the third control design approach between classical control and modern control. More specifically, it is an algebraic design approach on polynomial ring, and not on rational functions. Because rational functions are carefully avoided, differential equations are used directly. It is not necessary to use the Laplace transform.

In Chap. 1, we introduce the basic concept of CDM through a simple controller design example. Especially, "Sect. 1.2 A Simple Design Problem" will give the reader an overview of CDM design clearly. It describes the system representation used in CDM, the outline of the design process, and the control structure. Chapters 2 and 3 explain the basic theory and design method of CDM. In Chap. 4, we present four selected examples applying to advanced practical control design problems: a simple controller for an inverted pendulum mounted on a toy car, vibration suppression control for two-inertia system combined with a spring, a solution to the ACC benchmark problem where a two-mass-spring system with parameter uncertainty as well as the design constraints is considered, and a longitudinal control of a modern aircraft as a robust MIMO control design problem. In this approach, we need to calculate algebraic equations associated with the Diophantine equation and draw polynomial coefficients curves on the coefficient diagram. A CDM Toolbox for use with MATLAB for this purpose was developed and is given in the Appendix.

This book is not intended to be a standard textbook for control education. Rather it is intended to supplement those textbooks at the design/synthesis stage. The author used the textbooks by Franklin [3] and Chen [4], and some of the examples are designed by CDM in comparison. To learn CDM, basic mathematics such as algebra is necessary. However, such mathematics as the Laplace transform or matrix algebra are carefully avoided, because such mathematics will give much burden to chemical or biological engineers who are interested in control.

S. Manabe wishes to express his sincere thanks to many people who supported his research on many occasions in the last fifty years. Professors Warren and Weimer at the Ohio State University introduced the author at an early age to "Automatic Control". Messrs. M. Yokosuka, H. Takeda, H. Morikawa, and N. Mitani helped the author to pursue the control research at Mitsubishi Electric. Professor A. G. J. MacFarlane of Cambridge University gave the author valuable suggestions about the meanings of the works of J. C. Maxwell and E. J. Routh. These works constitute the basis of CDM, and his suggestions helped the author to reorient the course of his research. Professor Y. Hori of Tokyo University and members of the Motion Control Committee (Institute of Electrical Engineers of Japan) were keenly interested in CDM and gave valuable advice on various occasions. Professors Y. Nozaka and M. Iida of Tokai University helped the author on many occasions. The students and colleagues of Tokai University helped in the development of CDM through lively discussion in the classroom and laboratory. Professor Young-Chol Kim of Chungbuk National University of Korea and his students had a keen interest in CDM. They made pioneering efforts to make CDM known in the international community. Without their help and efforts, CDM would not have come to this present stage. S. Manabe also wishes to express his sincere gratitude to his wife, Yasuko Manabe, for her spiritual and physical support especially at the advanced age.

Y. C. Kim would like to express his utmost respect and gratitude to Dr. Manabe. Since his first meeting at ASCC in 1997, Dr. Manabe has taught him CDM through numerous discussions and workshops and has also worked on this subject together. It is a great pleasure to express his gratitude to his gurus L. H Keel and S. P. Bhattacharyya for their friendships, encouragement, and valuable teachings. Dr. Kim wants to express his deep gratitude to his beloved wife Agnes Jaesook Kim for her love, patience, and support.

This book was first intended as a full text including more advanced contents. However, because there is a lot of interest in CDM recently, but there are health problems to the authors, they decided to publish the first part of the book. The final edition will be published in the later years. They hope this book will help make future textbooks in the field of control much easier to understand. This book inevitably has errors, and we welcome corrective feedback from the readers. We also apologize in advance for any omissions or inaccuracies in referencing and would want to compensate for them in the final edition.

Fujisawa, Japan — Shunji Manabe
Cheongju, Korea (Republic of) — Young Chol Kim

References

1. Kessler C (1960) Ein Beitrag zur Theorie mehrschleifiger Regelungen. Regelungstechnik 8 (8):261–266
2. Lipatov AV, Sokolov NI (1978) Some sufficient conditions for stability and instability of continuous linear stationary systems. Translated from, Automatika i Telemekhanika 9:30–37, 1978; *Automat. Remote Contr.*(1979) 39:1285–1291
3. Franklin CF, Powell JD, Emami-Naeini A (2015) Feedback control of dynamic systems (7th edition). Pearson Education Ltd., Essex, England
4. Chen CT (1993) Analog and digital control system design: Transfer-function, state-space, and algebraic methods. Saunders College Publishing, Orlando

Contents

Chapter 1
Introduction

Abstract The overall picture of the Coefficient Diagram Method (CDM) is explained in this chapter. In **Basic Philosophy**, the basic philosophy of the author in writing this book is briefly introduced. In **Simple Design Problem**, the overall design procedure of CDM is shown by the design example of a simple position control system. In **System Representation**, the polynomial expression used in CDM is compared with the transfer function expression in classical control and the state-space expression in modern control. In **Outline of Design Process**, the design steps in CDM are explained in the order of design. In **Control Structure**, it is shown that various forms of controllers derived from various control theories can be represented by equivalent CDM controllers. In **Historical Background**, the history of CDM in the last 50 years is looked back. In **Summary**, important points are summarized and the six features of CDM are explained.

1.1 Basic Philosophy

In the beginning, the author wishes to express his basic philosophy through three topics; namely control, feedback control, and algebraic approach. "**Control is compromise**" is the basic philosophy of this book. The background of this philosophy is first explained. Then, the structure of control in a broad sense is presented, and it is shown that **Feedback control is only a small part of control**. Finally, **Algebraic Approach** is briefly explained and compared with classical control and modern control. It is shown that CDM, a specific type of algebraic approach, is developed as an answer to various design issues.

"**Control is compromise**." is the basic philosophy of this book. *Control* in Chinese character is composed of two characters; the first character means *trimming tree* and the second character means *riding horse*. You can trim branches of trees, but you cannot grow the branches on the tree. You have to wait until the branch grows again. In riding a horse, the horse cannot run as fast as 100 km/h or as slow as 1 km/h. He

S. Manabe and Y. C. Kim, *Coefficient Diagram Method for Control System Design*,
Intelligent Systems, Control and Automation: Science and Engineering 99,
https://doi.org/10.1007/978-981-16-0546-8_1

has a characteristic speed, which he likes best. The rider has to be satisfied with a compromise between his desire and the desire of the horse.

History shows that the word *Control* appeared first in *Records of Historian* written two thousand years ago by Szuma Chien in the Han Dynasty. In *The First Emperor of Chin*, the second emperor of Chin made the following remark as a reply to the advice of his wise ministers that the heavy tax should be reduced in order to keep peace in the country.

> But what is splendid about possessing an empire is being able to do as you please and satisfy your desires. By stressing and clarifying the laws, a ruler can stop his subjects from doing evil and so *control* the land within the seas.

The second emperor used *control* as controlling his subjects as he would. Such attitude is completely against the traditional wisdom for good government, and it is very natural that the dynasty ended very soon with the third emperor.

Good control is the result of a compromise between the desire of the controller and that of the object to be controlled. Or it is the compromise between *what should be done* and *what can be done*. The philosophies of present control design/synthesis theories are located somewhere in the spectrum of this compromise; classical control is more or less at *can be done* end, while modern control at *should be done* end.

In modern control, too much emphasis is placed on optimality. In a world where compromise is important, optimality is not a good philosophy. For one thing, in the process of seeking optimality, some kind of performance index is introduced. The effort to attain such optimality tends to give an answer, which is optimal only in that index and very poor in the other index. The common sense tells that if optimization is stopped at a certain point below the optimum, performance is fairly good even at the other index. It can be said that the second best is also the second best even at the other performance index. For another thing, the optimum is usually not a peak, but rather a range or plateau. Thus, the optimum cannot be obtained by mathematics and is chosen by the common sense of the designer. For these reasons, optimality is avoided as far as possible in this book, and compromise is more stressed.

Because compromise is very important in control system design, some mechanism to facilitate compromise has to be imbedded in the design process. Human being has a strong capacity of compromise when the problem is expressed in a graphical form. A graphical representation is a key to compromise, and serious attention has been paid to it. *Graph rather than mathematics* is also a feature of this book.

Feedback control is only a small part of control. This book is intended for the design of feedback control of dynamic systems. The main topic is on Linear Time Invariant (LTI) system. Although feedback control is very important, it is only a small part of total control. The total control system is shown in Fig. 1.1. It consists of a plant, feedback controller, and intelligent controller.

Input to the plant is applied through actuators, and the output of the plant is measured by sensors. The plant is a dynamic system, which can be expressed by differential equations. The plant considered here is an energy device in the sense that its performance is limited by energy. The feedback controller is an information device in the sense that sensor outputs are connected to controller inputs as information and

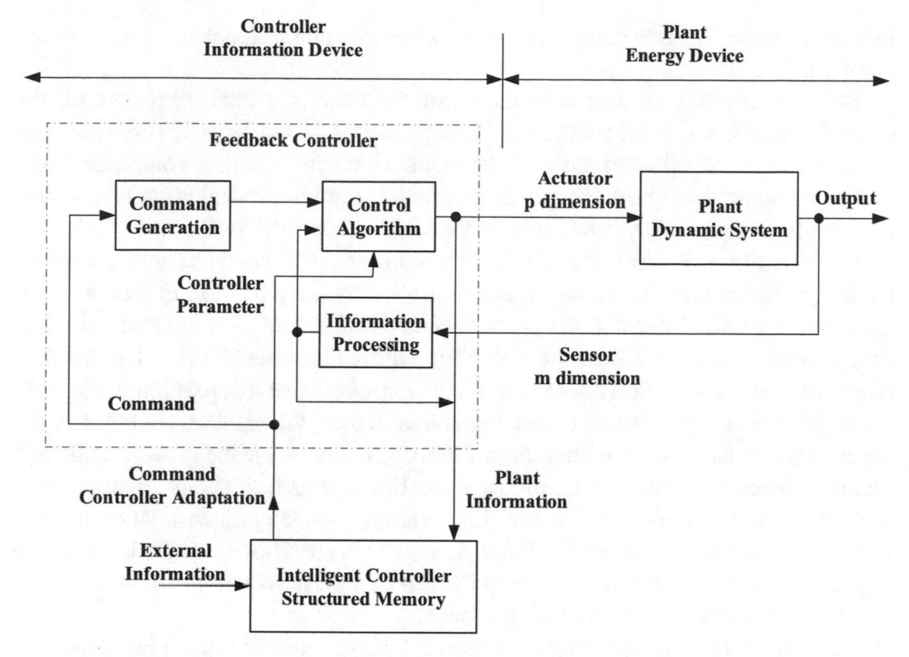

Fig. 1.1 Control system

controller outputs are fed to actuator inputs in the form of information. Information is usually in electronic form. The feedback controller has three functions; namely command generation, control algorithm, and sensor information processing.

This feedback controller is connected to the intelligent controller, which is also an information device and consists of large structured memory, which stores many procedures and large past experiences. The main purpose of the intelligent controller is to give proper instruction to the feedback controller as to what kind of control is to be performed at the specific moment. Such instruction is usually composed of one for command generation and the other for modification of control algorithm. The input to the intelligent controller is the plant information obtained from the actuator and sensor information. Some exterior information is fed to the intelligent controller as the basic command of overall control. Usually, many feedback controllers are under the control of an intelligent controller. The intelligent controller is placed at a higher layer than feedback controllers in the hierarchy.

The feedback controller and the intelligent controller constitute the controller in a broad sense. Usually, they are in a control computer. In the olden age before the control computer was introduced, the feedback controller was an analog type, and the relay sequence circuit performs the function of the intelligent controller. It is interesting to note that, in the actual system, the majority of the control function is devoted to the intelligent control and only a small portion, 5 to 10%, is used for feedback control. This means that feedback control is only a small part of control in a broad sense. For this reason, too much sophistication of feedback control is not justified.

Because feedback control is in the hierarchy under the intelligent control, the controller can be adapted to plant condition, and it can be provided with some measure to cope with the nonlinear nature of the plant. Thus, the feedback controller in the LTI environment must be as simple as possible. Otherwise, the addition of adaptive and nonlinear function by intelligent control will become difficult.

In olden times, controller and plant are both energy devices as seen in Watt's fly-ball governor and the steam engine controlled by the governor. In recent years, the plant may be an information system. Then controller and plant are both information devices. The energy limitation problem will not become evident. This book is concerned with a specific system where the controller is an information device and the plant is an energy device. Under this circumstance, the actuator usually has the severest power and response limitation. These limitations have the greatest influence on the performance of the closed-loop system. Because such actuators are expensive, the number of actuators, expressed as dimension p, is usually much smaller than the number of sensors, expressed as dimension m. This specific circumstance tends to make the system in the form of a loop instead of a mesh, where the actuator plays the role of a common node for various loops.

This total control system may be compared with control of human body motion. The human body is the plant. The actuators are feet and hands. The sensors are eyes, ears, skin, sense of muscle force, sense of balance, etc. The feedback controller may be the cerebellum and brain stem, and the intelligent controller may be the cerebrum. The human body is an energy device, and the performance is limited by the energy limitation. The number of actuators is much smaller than that of sensors. The vast memory accumulated in the cerebrum through the past education and experience works as the intelligent controller with much flexibility and complexity.

Algebraic Approach lies between classical control and modern control as shown in Fig. 1.2. It is the third control theory. It is often called the polynomial approach because polynomials are used in system representation. In classical control, frequency response design and root locus design are currently used. Both methods use transfer function in the system representation. The relation of controller parameters and closed-loop characteristics is graphical; the Bode/Nyquist diagram for frequency response design, and the root locus diagram for root locus design. In the design process, the basic controller structure with undetermined parameters is first assumed, and such parameters are adjusted so that the closed-loop system meets design specifications. Such design process is called as *Outward Approach* [1].

In modern control, pole assignment, optimal control (LQR, LQG), and H-infinity design are currently used. They all use state space in system representation. The relation of controller parameters and closed-loop characteristics is expressed in equations. In pole assignment, pole location representing closed-loop characteristics is related to controller parameters in algebraic matrix relation such as Ackermann's formula [2]. In optimal control and H-infinity, closed-loop characteristics are related to controller parameters through the Riccati equation. In the design process, an overall closed-loop system, to meet design specifications, is first chosen, and the controller

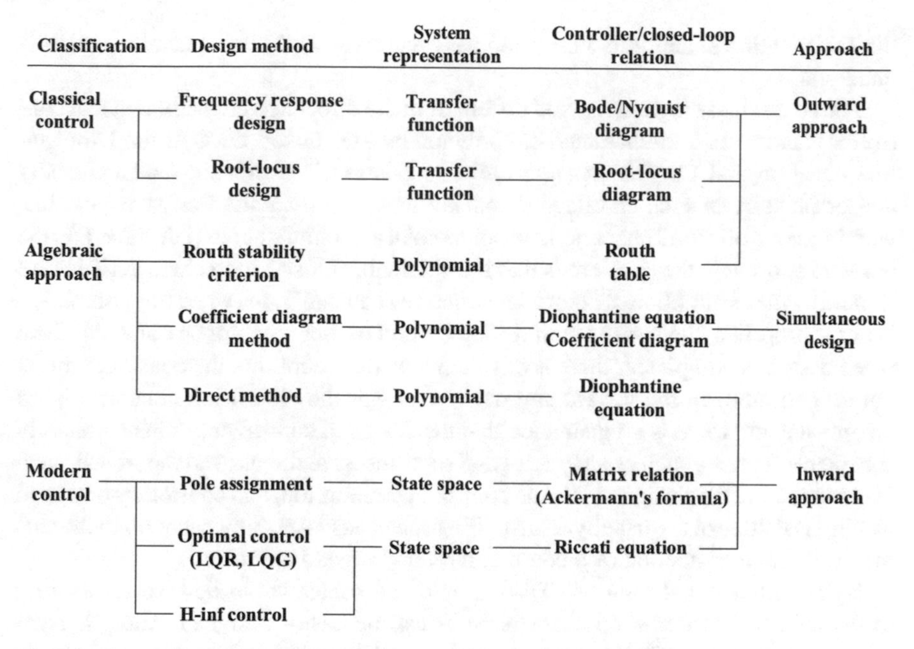

Fig. 1.2 Comparison of control theories

is obtained as the solution of equations. Such design process is called as *Inward Approach* [1].

Contrary to classical and modern control, the algebraic or polynomial approach is not clearly defined, because it is at the developing stage. In a broad sense, it includes the design approach by Routh stability criterion. It uses polynomial as system representation, and controller parameters are related to closed-loop stability through the Routh table. The design process is definitely an outward approach. In a narrow sense, the algebraic design is characterized by the direct design method [2]. The system representation is by the polynomial. The closed-loop characteristics are represented by the characteristic polynomial. It is related to the controller parameters through the Diophantine equation. In the usual design process, the characteristic polynomial is chosen first, and the controller is obtained as the solution of the Diophantine equation. Thus, it is definitely an inward approach. The problem is extended to the MIMO case by extension of the polynomial to the polynomial matrix. Various researchers have made intensive study already [1–4], but no serious study has been made for the selection of the characteristic polynomial. The characteristic polynomial is usually specified by pole assignment.

The transfer function expression is easy to understand but is inaccurate at pole-zero cancellation, while the state-space expression is accurate but difficult to understand. The polynomial expression is easy to understand like the transfer function, while it is accurate as the state-space expression. This is an advantage of the algebraic approach over classical control and modern control. The other advantage is that the Diophantine equation is linear and easy to solve, while the Riccati equation is quadratic.

The coefficient diagram method differs from the direct design method in that controller parameters and characteristic polynomial are related through the Diophantine equation and *Coefficient Diagram*. The relation is expressed mathematically and graphically in such an effective manner that *Simultaneous Design* is possible, whereby controller and characteristic polynomial are simultaneously designed. In the outward approach, the problem is that the obtainable closed-loop characteristics are limited by the assumption of controller structure at outset. Thus, when the assumption is not appropriate, the desired closed-loop characteristics may not be obtained. Even if the design is completed, there always remains the doubt that the controller might not be optimum. In the inward approach, although the closed-loop characteristics are guaranteed, there is no guarantee that the designed controller retains commonly acceptable features such as simplicity and robustness. In the outward approach, such features are usually designed into the controller assumption, and controllers designed by classical control are usually robust. The robustness issue only came up to the surface with the introduction of modern control and inward approach.

By simultaneous design of CDM, simple and robust controllers corresponding to the specified closed-loop characteristics can be easily designed. The graphical expression of the coefficient diagram makes intuitive design possible as in classical control with the Bode/Nyquist diagram. In control design, the simplicity of the controller and closed-loop performance are trade-off issues. The controller represents *what can be done*, while the closed-loop performance represents *what should be done*. To find a good compromise between the two is the key to good control. Compromise is the most important in control design, and CDM gives some answers to it.

1.2 A Simple Design Problem

In order to show the general picture of the Coeffcient Diagram Method, a simple design example of a position control system is presented in this section. Five topics, namely **Problem Statement**, **Definition of Stability Index and Equivalent Time Constant**, **Design Process**, **Characteristics of Control System**, and **Coefficient Shaping** will be represented in order.

Problem Statement will be first made. The system considered is a generic position control system shown in Fig. 1.3. The plant consists of a power amplifier and a motor. It can be represented in the differential equation form as

$$(0.25s + 1)(s + 1)sy = u, \tag{1.1}$$

$$v = sy, \tag{1.2}$$

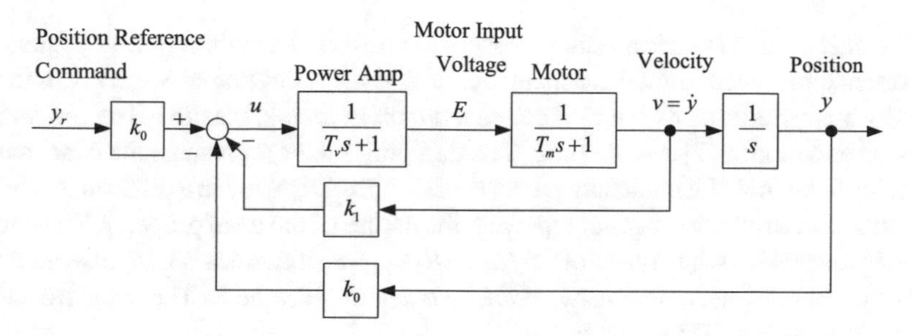

Fig. 1.3 Position control system

where u is the input to the power amplifier; v the velocity; y the position. The s stands for differential operator d/dt. The controller is a PD controller, where the velocity and position sensors obtain the necessary feedback signals. It can be represented by the differential equation form as

$$u = k_0(y_r - y) - k_1 v, \tag{1.3}$$

where y_r is the position reference command; k_0 position gain; k_1 velocity gain. By eliminating u from these equations, the equation which relates y to y_r is obtained as

$$(0.25s^3 + 1.25s^2 + s + k_1 s + k_0)y = k_0 y_r. \tag{1.4}$$

The term preceding y of the left side is the characteristic polynomial $P(s)$.

$$P(s) = 0.25s^3 + 1.25s^2 + (1 + k_1)s + k_0. \tag{1.5}$$

The problem is to find the controller gains k_1 and k_0 such that the system has good characteristics in terms of stability, response, and robustness

Definition of Stability Index and Equivalent Time Constant is next made as preparation for design. The characteristic polynomial is generally expressed in the following form.

$$P(s) = a_n s + a_{n-1}s^{n-1} + \cdots + a_1 s + a_0 = \sum_{i=0}^{n} a_i s^i. \tag{1.6}$$

The stability index γ_i and the equivalent time constant τ are defined as follows:

$$\gamma_i = a_i^2/(a_{i+1}a_{i-1}), \quad i = 1, \cdots, n-1. \tag{1.7}$$

$$\tau = a_1/a_0. \tag{1.8}$$

The stability index is an important measure to indicate the stability of the system. Its importance has been known for many years, and different names were given to this term by many authors. Kessler [5] called it *damping factor*; Naslin [6] *characteristic ratio*; Brandenburg [7] *double ratio*. The damping factor represents the exact nature of the term, but may be misleading, because it is currently used in a different meaning. The characteristic ratio may not represent the nature of the term properly. The double ratio represents how it is made of $(a_i/a_{i-1})/(a_{i+1}/a_i)$, but does not represent what it is. For these reasons, a new name *stability index* is given here. The standard values recommended in CDM are as follows:

$$\gamma_{n-1} = \cdots = \gamma_3 = \gamma_2 = 2, \qquad \gamma_1 = 2.5. \tag{1.9}$$

The reason why these values are recommended is the key to CDM and will be made clear in later chapters. The *equivalent time constant* is a measure of the response speed. When γ_i takes the standard values, all transient response settles within 2.5–3 times of τ. CDM design is based on stability index and equivalent time constant.

Design Process will be shown next. Because the problem is simple, the standard values of γ_i ($\gamma_2 = 2$, $\gamma_1 = 2.5$) can be chosen. By the definition of stability index, the following relations are derived.

$$a_1 = 1 + k_1 = a_2^2/(a_3\gamma_2) = 1.25^2/(0.25 \times 2) = 3.125, \tag{1.10}$$

$$a_0 = k_0 = a_1^2/(a_2\gamma_1) = 3.125^2/(1.25 \times 2.5) = 3.125. \tag{1.11}$$

Then, the design results are as follows:

$$k_1 = a_1 - 1 = 2.125, \qquad k_0 = 3.125, \qquad \tau = 1. \tag{1.12}$$

The resulting controller shows good characteristics, namely no-overshoot, fast response, and good robustness. The settling time is about 2.5–3 times of τ. It should be noted that simple arithmetic suffices for the design with no need for higher mathematics.

Under some circumstances, slower responses are permissible. Then design proceeds for $\gamma_2 > 2$ and $\gamma_1 = 2.5$ with $k_1 = 2.125 \sim 0$ as a free parameter.

$$k_0 = a_0 = a_1^2/(a_2\gamma_1) = (1 + k_1)^2/(1.25 \times 2.5) = 0.32(1 + k_1)^2, \tag{1.13}$$

$$\gamma_2 = a_2^2/(a_3a_1) = 1.25^2/[0.25(1 + k_1)] = 6.25/(1 + k_1) > 2, \tag{1.14}$$

$$\tau = a_1/a_0 = 3.125/(1 + k_1). \tag{1.15}$$

When k_1 becomes 0, the response is the slowest. Then $k_0 = 0.32$, $\gamma_2 = 6.25$, and $\tau = 3.125$. If further slow response is allowed, design proceeds for $k_1 = 0$, $\gamma_2 = 6.25$, and $\gamma_1 > 2.5$ with $k_0 = 0.32 \sim 0$ as a free parameter.

$$\gamma_1 = a_1^2/(a_2 a_0) = 1/(1.25k_0) > 2.5, \tag{1.16}$$

$$\tau = a_1/a_0 = 1/k_0 > 3.125. \tag{1.17}$$

The above design procedure shows that CDM design is simple and flexible in meeting the specific need of design. The effectiveness of CDM comes from the standard values of the stability indices into which experiences of past design efforts are crystallized.

Characteristics of Control System will be discussed hereafter. Table 1.1 shows the case when τ is increased from the designed nominal value 1 to 2, 3.125, and 5. These are the cases where slower responses are permissible. In order to retain favorable responses, the stability indices are increased in a systematic manner. The first step is to increase γ_2 with fixed $\gamma_1 = 2.5$ until k_1 becomes 0. The second step is to increase γ_1 with fixed $\gamma_2 = 6.25$ and $k_1 = 0$ by decrease of k_0. The coefficient diagram is shown in Fig. 1.4. In coefficient diagram, the coefficients a_i of the characteristic polynomial is shown in the ordinate in logarithmic scale, while order i is shown in abscissa in decreasing order. In the nominal case #1,

$$a_i = [a_3 \ \ a_2 \ \ a_1 \ \ a_0] = [0.25 \ \ 1.25 \ \ 3.125 \ \ 3.125]. \tag{1.18}$$

The step responses with different τs are shown in Fig. 1.5. All responses are smooth and show no-overshoot.

Table 1.2 shows the case when γ_1 is made one half and two times with k_0 made twice and one half. The coefficient diagram is shown in Fig. 1.6. The step responses are shown in Fig. 1.7. When γ_1 is small, the stability deteriorates and conspicuous overshoot is observed.

From these observations, we can see that the response speed is mainly affected by τ, and the waveform of the step response is largely influenced by γ_1. These two parameters seem to characterize the step response. The stability index γ_i is represented by the curvature of the coefficient diagram at the specific order i, as is clear from the definition of the stability index. The equivalent time constant is represented by the inclination of the diagram at the rightmost ends. The coefficient diagram and the step response are closely related through stability indices and equivalent time constant. If the coefficient diagram is given, the approximate step response can be visualized and vis-á-vis. This problem will be discussed in later chapters.

The above results are shown on the parameter space by k_0 and k_1 as in Fig. 1.8. From the Routh stability condition, $a_0 > 0$, $a_2 a_1 > a_3 a_0$, the stable region is given as follows:

$$0 < k_0 < 5(1 + k_1). \tag{1.19}$$

Because such a stable region is too broad, proper parameter selection cannot be done. When $\gamma_1 = a_1^2/(a_2 a_0)$ is specified as 2.5, and γ_2 and τ are allowed to change, the parameter satisfy the following equation.

$$k_0 = 0.32(1 + k_1)^2. \tag{1.20}$$

Table 1.1 τ variation

Case	k_1	k_0	τ	γ_2	γ_1
#1	2.125	3.125	1	2	2.5
#2	0.5625	0.78125	2	4	2.5
#3	0	0.32	3.125	6.25	2.5
#4	0	0.2	5	6.25	4

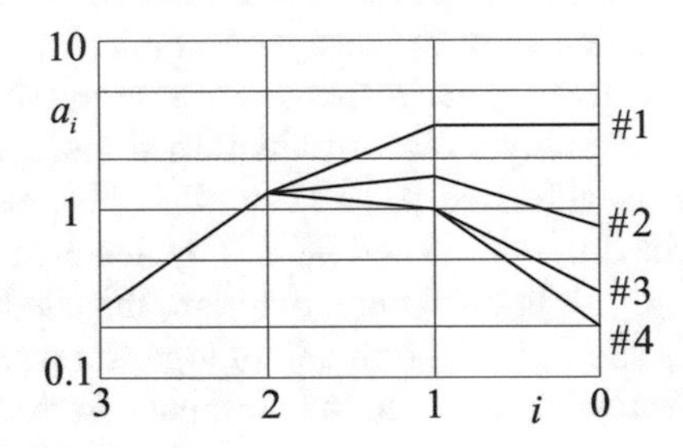

Fig. 1.4 Coefficient diagram for τ variation

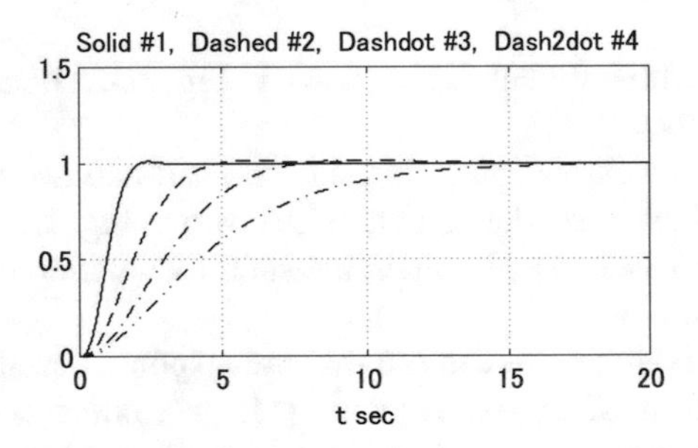

Fig. 1.5 Step responses to τ variation

Table 1.2 γ_1 variation

Case	k_1	k_0	τ	γ_2	γ_1
#1	2.125	3.125	1	2	2.5
#5	2.125	6.25	0.5	2	1.25
#6	2.125	1.5625	2	2	5

The responses are smooth on every point on the curve, oscillatory on the right side, corresponding to small γ_1, and sluggish on the left side. Although the parameters can be chosen at any point on the curve, the best choice is at point #1, where τ is smallest and the response is fastest.

Coefficient Shaping is a graphical design approach where the close relation between the coefficient diagram and the step response is fully utilized. By this approach, the characteristic polynomial and the controller can be simultaneously

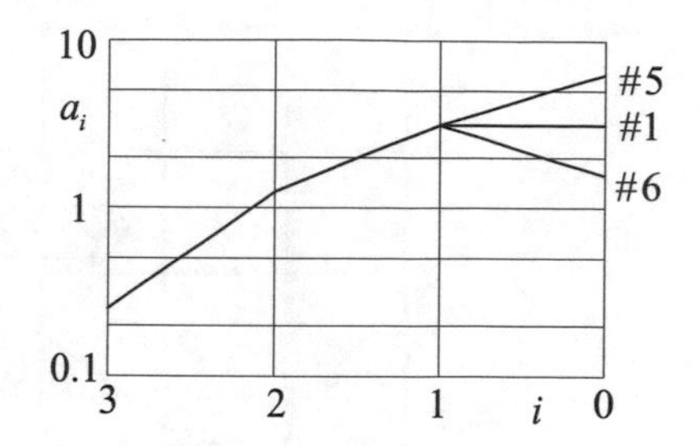

Fig. 1.6 Coefficient diagram for γ_1 variation

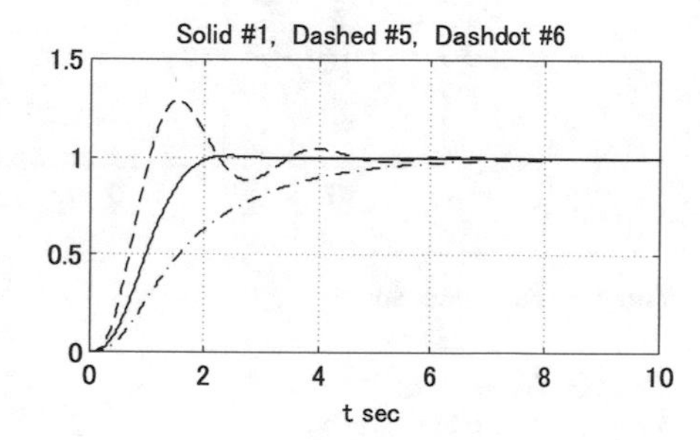

Fig. 1.7 Step responses to γ_1 variation

designed. This is the most conspicuous feature of CDM, and its implication will be discussed later. A brief description of this approach will be now shown. In this example, the characteristic polynomial is composed of two component polynomials.

$$\begin{aligned} P(s) &= P_0(s) + P_k(s), \\ P_0(s) &= 0.25s^3 + 1.25s^2 + s, \\ P_k(s) &= k_1 s + k_0, \end{aligned} \tag{1.21}$$

where $P_0(s)$ is the characteristic polynomial without controller, and $P_k(s)$ is the contribution by the controller. The coefficient diagram is shown in Fig. 1.9. The $P_0(s)$ is shown by small circles and dash-dot line. The $P_k(s)$ is shown by small square and dotted line. With this decomposed coefficient diagram, the variation of $P(s)$ due to parameter variation can be easily visualized. Because the general shape of $P(s)$ in the coefficient diagram is closely related to the step response, the variation of the step response can be easily estimated.

The first step in the design is to draw $P_0(s)$ in the coefficient diagram. We draw a rough sketch of $P(s)$ based on the required step response. The $P_k(s)$ is designed to fill the gap between $P(s)$ and $P_0(s)$. The final $P(s)$ is obtained as the sum of $P_0(s)$ and $P_k(s)$. Thus $P(s)$ and $P_k(s)$ are simultaneously designed. This approach is very effective in defining the basic control structure. The stability indices and equivalent time constant best fitted to the purpose can be easily estimated. Because this is a graphical approach, the intuition of the designer is fully utilized and design is more efficient.

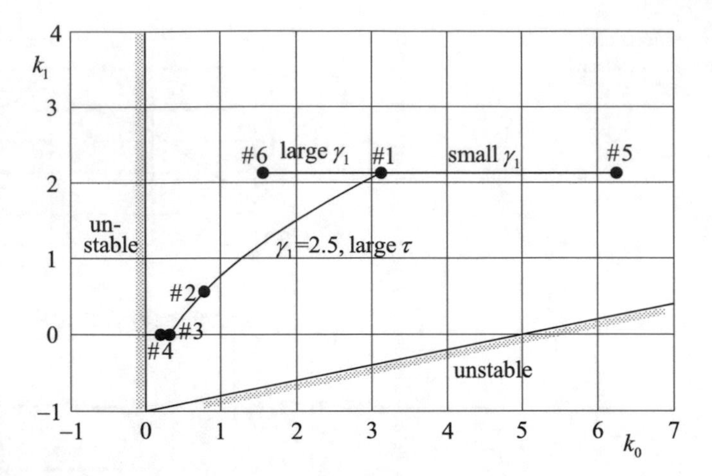

Fig. 1.8 Parameter space

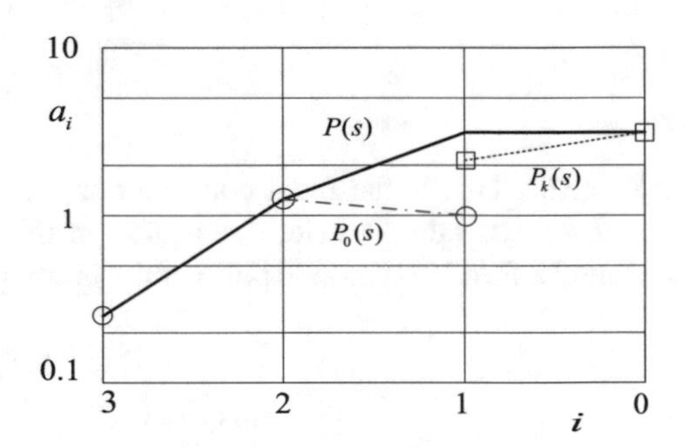

Fig. 1.9 Coefficient diagram, coefficient shaping

1.3 System Representation

In classical control, the transfer function is used for system representation. In modern control, the state space is used. These two approaches are quite different, and the difference seems to stem from the difference of expression in system representation [8, p. 471]. As any culture is characterized by the language used in the specific culture, it is natural that the problem to be handled and the results of the design are greatly influenced by such difference of expression. The Coefficient Diagram Method is intended to be used by control non-specialists for the design of practical controllers in their specific field of application. For this reason, *Polynomial Expression* is used in CDM for the system representation. The polynomial expression is the same as the differential equation expression, commonly used for the analysis of the dynamic system.

In this section, **Polynomial Expression** used in CDM is first explained. Then, it is compared with **Transfer Function Representation** and **State-Space Representation**. It will be made clear that the polynomial expression enjoys the merits of these

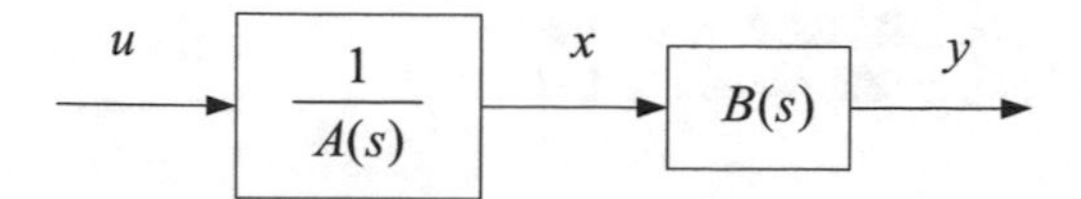

Fig. 1.10 Block diagram of polynomial expression, RPF

expressions; namely, the convenience of the transfer function and accuracy of state space.

Polynomial Expression used in CDM is directly derived from the differential equation of the dynamic system. The system may be expressed as follows:

$$a_3\frac{d^3y}{dt^3}+a_2\frac{d^2y}{dt^2}+a_1\frac{dy}{dt}+a_0y=b_2\frac{d^2u}{dt^2}+b_1\frac{du}{dt}+b_0u, \tag{1.22}$$

where u is the input, y is the output, and t is the time. Now the differential operator d/dt is expressed as s. Then, the equation becomes

$$A(s)y=B(s)u, \tag{1.23}$$

$$A(s)=a_3s^3+a_2s^2+a_1s+a_0, \tag{1.24}$$

$$B(s)=b_2s^2+b_1s+b_0. \tag{1.25}$$

In Eq. (1.23), new variable x, such as $y=B(s)x$, is introduced. Then, another expression is derived.

$$A(s)x=u, \qquad y=B(s)x. \tag{1.26}$$

In CDM, either Eq. (1.26) or Eq. (1.23) is used as the standard system representation. The former is called as *Right Polynomial Fraction* (RPF) and the latter is called as *Left Polynomial Fraction* (LPF).

Because the Laplace transform is not used at all in this development, the polynomials $A(s)$ and $B(s)$ are always prefixed before some variables. They are never placed in denominator. At this point, the polynomial expression used in CDM is different from the standard polynomial representation. In block diagram, some polynomial is placed in denominator as in Figs. 1.10 or 1.11. These are to be understood as representing Eqs. (1.26) or (1.23). Transfer function $G(s)$ is expressed as

$$y=G(s)u=\frac{B(s)}{A(s)}u. \tag{1.27}$$

This is to be understood as either Eq. (1.26) or Eq. (1.23). The ambiguity exists in transfer function expression.

Multi-Input-Multi-Output (MIMO) can be handled as Single-Input-Single-Output (SISO) case. The expression for *Right Polynomial Matrix Fraction* (RPMF) is

$$A(s)x=u, \qquad y=B(s)x, \tag{1.28}$$

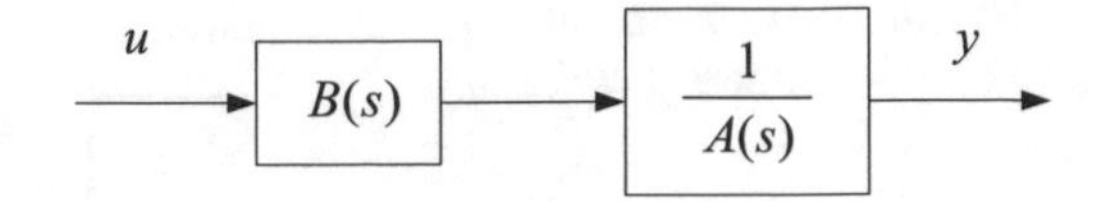

Fig. 1.11 Block diagram of polynomial expression, LPF

and the expression for *Left Polynomial Matrix Fraction* (LPMF) is

$$A^*(s)y = B^*(s)u, \tag{1.29}$$

where x, y, and u are vectors of proper dimensions, and $A(s)$, $B(s)$, $A^*(s)$, and $B^*(s)$ are matrices of proper dimensions. Because RPMF and LPMF are equivalent, the following relation must hold.

$$A^*(s)B(s) = B^*(s)A(s). \tag{1.30}$$

Because the commutative law is not applicable to matrices, $A^*(s)$ and $B^*(s)$ are different from $A(s)$ and $B(s)$.

Polynomials are *Ring* because only addition, subtraction, and multiplication are possible. Transfer functions are *Field* because addition, subtraction, multiplication, and division are possible. Because the Laplace transform is not used, the polynomial expression of the nonlinear differential equation is permissible. For example, dy^2/dt can be interpreted as either sy^2 or $2ysy$.

The differentiation in non-integer order (fractional order) such as $d^{1.4}y/dt^{1.4}$ will be represented as $s^{1.4}y$. The dynamic system expressed by a differential equation of non-integer order is currently called as *Fractional-Order System* (FOS). Extension to such a system can be made by simply allowing the non-integer order of s such as $s^{1.4}$ in the polynomials.

Transfer Function Expression is commonly used in classical control. The expression is compact and easy to handle. However, it has three drawbacks. The first drawback is that it is based on the Laplace transform. The Laplace transform is familiar to electrical engineers, but it is an obstacle to specialists in other fields, who may try to introduce control to their own dynamic system. Such specialists may be nonelectrical engineers such as physicists, chemists, biologists, and economists. Also, the Laplace transform presupposes the system is linear. The application of the Laplace transform to a nonlinear system is possible, but it is too cumbersome. The dynamic system in real world is usually nonlinear. The linear system is only an approximation. Exclusion of nonlinearity from the outset loses important ingredients of the real system.

The second drawback is the transfer function expression presupposes a zero-initial condition. Thus, the system analysis for non-zero initial condition is very cumbersome, although not impossible.

The third drawback is the ambiguity inherent in transfer function expression. Suppose the transfer function from u to y is given as follows:

$$Y(s)/U(s) = G(s) = s/s, \tag{1.31}$$

where $Y(s)$ is the Laplace transform of y and $U(s)$ that of u. In RPF, this may be interpreted as

$$sx = u, \quad y = sx. \tag{1.32}$$

This lead to

$$y = u. \tag{1.33}$$

The result is correct, because y is the differentiation of the integral of u. In LPF, this may be interpreted as

$$sy = su. \tag{1.34}$$

This leads to $s(y - u) = 0$, and finally to

$$y = u + constant. \tag{1.35}$$

This result is correct because Eq. (1.34) only specifies that the derivatives of y and u are equal. The above example is the generic nature of the transfer function and shows the ambiguity of the transfer function. When the numerator and denominator have a common factor, the common factor may be cancelled out or has to be retained depending on its interpretation. This ambiguity is commonly understood as *Pole-Zero Cancellation* problem.

State-Space Expression is commonly used in modern control. The state-space expression has a close relation with the polynomial expression. For deriving the state-space expression corresponding to RPF as in Eq. (1.26), the following state variables are defined.

$$x_2 = \ddot{x} = s^2 x, \quad x_1 = \dot{x} = sx, \quad x_0 = x. \tag{1.36}$$

The state-space expression with a vector $\mathrm{x} := [x_2 \ x_1 \ x_0]^T$ is

$$\dot{\mathrm{x}} = A\,\mathrm{x} + Bu, \quad y = C\,\mathrm{x} + Du, \tag{1.37}$$

$$A = \begin{bmatrix} -a_2/a_3 & -a_1/a_3 & -a_0/a_3 \\ 1 & 0 & 0 \\ 0 & 1 & 0 \end{bmatrix}, \quad B = \begin{bmatrix} 1/a_3 \\ 0 \\ 0 \end{bmatrix},$$

$$C = [b_2 \ b_1 \ b_0], \quad D = [0].$$

This state-space expression derived from RPF is *control canonical form*. The x in RPF is named as *basic state variable* because all the states of state-space expression are x and its derivatives.

The state-space expression corresponding to LPF expression, shown in Eqs. (1.23), (1.24), and (1.25), is derived in the following manner. First, these equations are combined as follows:

$$(a_3s^3 + a_2s^2 + a_1s + a_0)y = (b_2s^2 + b_1s + b_0)u. \tag{1.38}$$

This is rearraged as follows:

$$\frac{a_0}{a_3}y - \frac{b_0}{a_3}u + s\left[\frac{a_1}{a_3}y - \frac{b_1}{a_3}u + s\left(\frac{a_2}{a_3}y - \frac{b_2}{a_3}u + sy\right)\right] = 0. \tag{1.39}$$

Now y is chosen as state z_2. The term inside ( ) is chosen as z_1. The term inside [ ] is chosen as z_0. Then, the following relations are derived.

$$y = z_2, \tag{1.40}$$

$$z_1 = sz_2 + (a_2/a_3)z_2 - (b_2/a_3)u, \tag{1.41}$$

$$z_0 = sz_1 + (a_1/a_3)z_2 - (b_1/a_3)u, \tag{1.42}$$

$$0 = sz_0 + (a_0/a_3)z_2 - (b_0/a_3)u. \tag{1.43}$$

These lead to the state-space expression known as *observer canonical form*. Letting $\mathrm{z} := [z_2 \;\; z_1 \;\; z_0]^T$, we have

$$\dot{\mathrm{z}} = F\,\mathrm{z} + G\,u, \quad y = H\,\mathrm{z} + J\,u. \tag{1.44}$$

$$F = \begin{bmatrix} -a_2/a_3 & 1 & 0 \\ -a_1/a_3 & 0 & 1 \\ -a_0/a_3 & 0 & 0 \end{bmatrix}, \quad G = \begin{bmatrix} b_2/a_3 \\ b_1/a_3 \\ b_0/a_3 \end{bmatrix},$$

$$H = [1 \;\; 0 \;\; 0], \quad J = [0].$$

Also, these new states are expressed by x of RPF, when u is replaced by $A(s)x$ and $y = z_2$ by $B(s)x$ in Eqs. (1.40), (1.41), and (1.42), as follows:

$$\begin{bmatrix} z_2 \\ z_1 \\ z_0 \end{bmatrix} = \begin{bmatrix} b_2s^2 + b_1s + b_0 \\ (s + a_2/a_3)(b_1s + b_0) - \{(a_1/a_3)s + a_0/a_3\}b_2 \\ \{s^2 + (a_2/a_3)s + (a_1/a_3)\}b_0 - (a_0/a_3)(b_2s + b_1) \end{bmatrix} x. \tag{1.45}$$

This leads to the following state transformation relation.

$$\begin{bmatrix} z_2 \\ z_1 \\ z_0 \end{bmatrix} = \begin{bmatrix} b_2 & b_1 & b_0 \\ b_1 & b_0 + (a_2/a_3)b_1 - (a_1/a_3)b_2 & (a_2/a_3)b_0 - (a_0/a_3)b_2 \\ b_0 & (a_2/a_3)b_0 - (a_0/a_3)b_2 & (a_1/a_3)b_0 - (a_0/a_3)b_1 \end{bmatrix} \begin{bmatrix} x_2 \\ x_1 \\ x_0 \end{bmatrix}. \tag{1.46}$$

In this manner, state-space expressions are derived for RPF and LPF. The states are defined in terms of the basic state variable x. The relation of state transformation is derived. Polynomial expressions have one-to-one correspondence with the state-space expressions; RPF to control canonical form, and LPF to observer canonical form.

Thus, the state-space expression is accurate with no ambiguity. It is easily extended to a nonlinear system. MIMO system can be handled in the same manner as SISO. The non-zero initial condition can be handled easily. Thus, the state-space expression is well fitted to machine computation. But it is not considered well fitted to design purpose because of the following three drawbacks:

The first drawback is too much diversity in state-space expression. The number of elements of the matrix increases in the proportion of the square of dimension, while the number of the coefficients of polynomial increase only in the proportion of the order. In the state space, an infinite number of expressions are possible for the same input–output relation. In polynomial expression, only two expressions, RPF and LPF, are allowed. Thus, when the controller is designed, the designer can easily understand the meaning of the controller in polynomial expression, while it is almost impossible in state-space expression because of too many elements and too many equivalent expressions.

The second drawback is unnecessarily rigid expression. In state space, the order has to be defined. The expression is in monic. In usual design, the order and the coefficient of the highest order are not known at the beginning. They are constantly modified with the course of the design process. Thus, such rigidity is very cumbersome in design.

The third drawback is that the connection of systems, in series or parallel, is fairly difficult in state-space expression. In polynomial expression, it can be easily done by multiplication or addition of polynomials.

Polynomial expression retains all the merits of transfer function expression and state-space expression, but also circumvents all these difficulties. When it is compared with transfer function expression, the difference is only in handling the numerator and denominator separately. In this way, the easiness of handling found in transfer function expression is inherited, while its drawbacks, such as the necessity of the Laplace transform, zero-initial condition presumption, and ambiguity, are avoided. When it is compared with state-space expression, the main difference is that only control canonical form and observer canonical form are allowed to use. In this way, the accuracy found in state-space expression is inherited, while its drawbacks, such as excess diversity, unnecessary rigidity, and difficulty in series–parallel connection, are avoided.

1.4 Outline of Design Process

The explanation of the design process by CDM is the topic of the entire book. Because the design process is quite different depending on the nature of the plant and the control requirement, meaningful explanations can be made only for specific

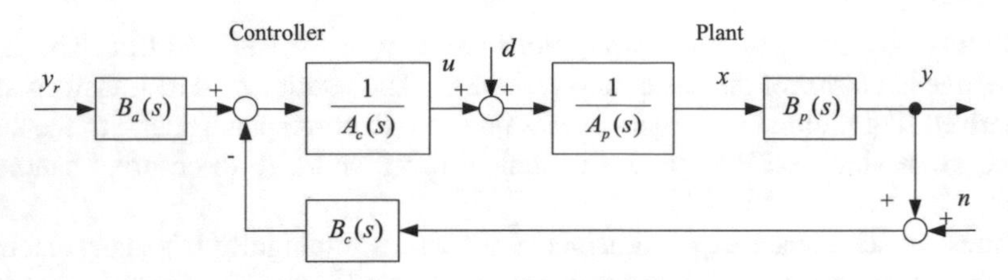

Fig. 1.12 Block diagram of a CDM standard feedback system

examples as will be shown in later chapters. In this section, only the outline of the design process is briefly explained. **Mathematical Model** used in CDM is first explained. Then, the **Input–Output Relations for CDM** are sought, and the meaning of such relations are discussed. Especially, the problem of **Pole-Zero Cancellation** will be discussed in some detail. Finally, **Design Steps in CDM** will be explained in general terms.

Mathematical Model used in CDM will be explained hereafter. The standard block diagram of CDM design for a Single-Input-Single-Output (SISO) system is shown in Fig. 1.12. The extension to Multi-Input-Multi-Output (MIMO) can be made with the proper interpretation, but it is not discussed here for simplicity. The plant equation is given as

$$A_p(s)x = u + d, \quad y = B_p(s)x, \tag{1.47}$$

where u, y, and d are input, output, and disturbance. The symbol x is called the basic state variable. $A_p(s)$ and $B_p(s)$ are the denominator and numerator polynomial of the plant transfer function $G_p(s)$. This expression is in RPF and corresponds to the control canonical form of the state-space expression. In the expression, all states are expressed by the basic state variable x and its high-order derivatives.

Controller equation is given as

$$A_c(s)u = B_a(s)y_r - B_c(s)(y + n), \tag{1.48}$$

where y_r and n are reference input and noise on the output. $A_c(s)$ is the denominator of the controller transfer function. $B_a(s)$ and $B_c(s)$ are called the reference numerator and feedback numerator of the controller transfer function. Because the controller transfer function has two numerators, it is called two-degree-of-freedom system. The transfer function $G_c(s) = B_c(s)/A_c(s)$ is the controller transfer function for feedback signal, and the transfer function $G_a(s) = B_a(s)/A_c(s)$ is the controller transfer function for reference signal. This expression is in LPF and corresponds to the observer canonical form of the state-space expression.

Eliminating the variables y and u from Eqs. (1.48) and (1.47) yields

$$P(s)x = B_a(s)y_r + A_c(s)d - B_c(s)n, \tag{1.49}$$

where $P(s)$ is the characteristic polynomial and given as

$$P(s) = A_c(s)A_p(s) + B_c(s)B_p(s). \tag{1.50}$$

In the mathematical model of CDM design, the plant is expressed in RPF and controller in LPF. Also, the controller is in a two-degree-of-freedom system. These choices make the design simple and consistent as will be clear in later chapters. The disturbance d is placed only to the plant input. Any disturbances placed to other parts of the plant have to be replaced by the equivalent disturbance placed at the plant input. In a similar manner, noise n is placed only to the plant output. Any noises placed at other parts have to be replaced by the equivalent noise.

Input–Output Relations for CDM will be derived from Eq. (1.49). In CDM, the response of the basic state variable x is considered to be the most important. Plant output y and input u are expressed in terms of x as follows:

$$y = B_p(s)x, \quad u = A_p(s)x - d. \tag{1.51}$$

Thus, the input–output relations are summarized as follows;

$$P(s)x = B_a(s)y_r + A_c(s)d - B_c(s)n, \tag{1.52}$$

$$P(s)y = B_p(s)[B_a(s)y_r + A_c(s)d - B_c(s)n], \tag{1.53}$$

$$P(s)u = A_p(s)[B_a(s)y_r - B_c(s)n] - B_c(s)B_p(s)d. \tag{1.54}$$

Because this system has three inputs (y_r, d, n) and three outputs (x, y, u), there are nine transfer functions. However, the polynomials to be designed are only four $(P(s), A_c(s), B_c(s), B_a(s))$, and only four relations will be sufficient for design. In CDM design, the following four basic relations are selected as standard.

$$P(s)x = P(0)y_r, \tag{1.55}$$

$$P(s)y = B_p(s)B_a(s)y_r, \tag{1.56}$$

$$P(s)y = B_p(s)A_c(s)d, \tag{1.57}$$

$$P(s)(-y) = B_p(s)B_c(s)n. \tag{1.58}$$

Equation (1.55) is the response of x to y_r when $B_a(s) = P(0)$, where $P(0)$ is the zeroth-order coefficient of $P(s)$, a_0. The transfer function corresponding to this relation, $T_0(s)$, is named as the zeroth-order canonical transfer function of $P(s)$, as will be explained later. This transfer function specifies the characteristic polynomial, and it is a very good measure of stability. Equation (1.56) is for the command following characteristics, and the corresponding transfer function is designated as $W(s)$. Equation (1.57) is for disturbance rejection characteristics, and the corresponding transfer function is designated as $T_{yd}(s)$. Equation (1.58) is for noise attenuation characteristics, and the corresponding transfer function is the complementary sensitivity function $T(s)$. This relation is useful for checking the robustness.

In the CDM design, these four basic relations are used as performance specification. The design of $P(s)$ is first made to satisfy specifications on Eqs. (1.55), (1.57), and (1.58), and then $B_a(s)$ is adjusted to satisfy the specification on Eq. (1.56). Thus disturbance rejection characteristics and command following characteristics can be specified independently. The design of $P(s)$ is feedback control design and that of $B_a(s)$ is feedforward control design.

In actual design, the specifications may be expressed in other forms. Then, such specifications must be interpreted to the above four relations before CDM design starts. By so doing, ambiguity and contradiction of specifications become evident and they are corrected automatically.

Pole-Zero Cancellation technique is commonly used in the traditional design using transfer function. When this problem is studied more carefully, the importance of the four basic relations for performance specification will become clear. The denominator polynomial $A_{cp}(s)$ and numerator polynomial $B_{cp}(s)$ of open-loop transfer function $G(s)$ for Fig. 1.12 are expressed as follows:

$$A_{cp}(s) = A_c(s)A_p(s), \quad B_{cp}(s) = B_c(s)B_p(s). \tag{1.59}$$

Then, the following relations are derived.

$$G(s) = B_{cp}(s)/A_{cp}(s), \tag{1.60}$$

$$P(s) = A_{cp}(s) + B_{cp}(s), \tag{1.61}$$

$$T(s) = B_{cp}(s)/P(s), \tag{1.62}$$

$$S(s) = A_{cp}(s)/P(s), \tag{1.63}$$

where $S(s)$ is known as sensitivity function. Pole-zero cancellation occurs when $B_c(s)$ and $A_p(s)$ have a common factor. Zero-pole cancellation occurs when $A_c(s)$ and $B_p(s)$ have a common factor. In both cases, $A_{cp}(s)$ and $B_{cp}(s)$ have a common factor. In such cases, the information for the common factor is invisible for the design using $G(s)$ as in classical control and for the design using $T(s)$ and $S(s)$ as in mixed-sensitivity design of H_∞. However, such common factor stays in $P(s)$, and the effects of such common factor become evident for some other input–output relations or in transient responses for some initial conditions. It is well known that, if such common factor is unstable, the system becomes unstable. But it is less known that when the real part is small, such as in oscillatory poles and zeros, even stable common factor causes deterioration of robustness and leads to instability, because there is always some error in such cancellation.

In CDM, usually design proceeds without pole-zero cancellation technique, because $P(s)$ is the target of design. Even when the technique is used, whether such pole-zero cancel is acceptable or not can be easily checked by $P(s)$, because unacceptable pole-zero cancellation makes some stability index very small.

Design Steps in CDM will be briefly explained hereafter. Before the design starts, the analysis of specifications must be done. Usually, specifications are vague and inconsistent. After the analysis, these specifications are interpreted to the specifications of four basic relations ($T_0(s)$, $W(s)$, $T_{yd}(s)$, and $T(s)$). They are further interpreted to the specifications of four polynomials ($P(s)$, $A_c(s)$, $B_c(s)$, and $B_a(s)$). Especially, the allowable range of stability indices and equivalent time constant concerning $P(s)$ are most important. In this way, specifications become clear and consistent.

The first step is to express the plant in RPF, where the denominator and numerator polynomials are as follows:

$$A_p(s) = d_{n_p}s^{n_p} + \cdots + d_1 s + d_0, \tag{1.64}$$

$$B_p(s) = n_{m_p}s^{m_p} + \cdots + n_1 s + n_0, \tag{1.65}$$

where d_i and n_i are the coefficients of these polynomials, and n_p and m_p are their orders. In CDM, the order of such plant is called as m_p/n_p order.

The second step is to express the controller in LPF and define the control structure. The control structure means the order of controller and allowable range of controller parameters. The denominator and numerator of controller are expressed as follows:

$$A_c(s) = l_{n_c}s^{n_c} + \cdots + l_1 s + l_0, \tag{1.66}$$

$$B_c(s) = k_{m_c}s^{m_c} + \cdots + k_1 s + k_0, \tag{1.67}$$

$$B_a(s) = m_{m_a}s^{m_a} + \cdots + m_1 s + m_0, \tag{1.68}$$

where l_1, k_i, and m_i are the coefficients of these polynomials, and n_c, m_c, and m_a are their orders. In CDM, the order of such controller is called as m_c/n_c order. Because reference numerator $B_a(s)$ has no effect on $P(s)$, its order is not referred. When the type of controller is known beforehand, the control structure is automatically derived. As an example, the PD controller introduced in Sect. 1.2 has the following control structure.

$$n_c = 0, \quad m_c = 1, \quad m_a = 0, \quad l_0 = 1, \quad k_1 > 0, \quad k_0 = m_0 > 0. \tag{1.69}$$

The PD controller is a 1/0 order controller. A PID controller has the following control structure:

$$n_c = 1, \quad m_c = 2, \quad m_a = 2,$$

$$l_1 = 1, \quad l_0 = 0, \quad k_2 = m_2 > 0, \quad k_1 = m_1 > 0, \quad k_0 = m_0 > 0. \tag{1.70}$$

PID controller is 2/1 order controller.

Usually, controller types are known beforehand by experience. But for a new design, the control structure has to be defined with careful consideration. It is well known that the order of the controller is affected by the order of the plant. But in reality, it is more influenced by the nature of the plant. Some plant is easy to control even when the order is high, and a simple PI controller (1/1 order) suffices. Some plant is difficult to control even when the order is low, and the controller order becomes close to the plant order. In a very difficult plant, such controller becomes non-minimum phase or unstable, and robustness can be greatly reduced. Under this circumstance, the stability and robustness of the closed-loop system become trade-off issues, and one has to be sacrificed for the other. The controller order is also affected by the nature of the actuators and sensors,too. Because actuators and sensors are considered as portions of the plant, their proper choice makes the nature of the plant more amenable to control.

At control structure design, the specifications concerning the low-frequency response of disturbance rejection, $T_{yd}(s)$, and high-frequency response of complementary sensitivity function, $T(s)$, are most instructive. Because control structure design is a complicated process, its complete treatment will be left for later chapters. Also, design examples will help to clarify the specific procedure.

The third step is to express the relation between the denominator/numerator polynomials of plant/controller and the characteristic polynomial. The relation is shown in Eq. (1.50) and is repeated here.

$$A_c(s)A_p(s) + B_c(s)B_p(s) = P(s). \tag{1.71}$$

This equation is commonly called as the Diophantine equation. Using Eqs. (1.66) and (1.67), this is expanded to the following form.

$$\left(\sum_{i=0}^{n_c} l_i s^i\right) A_p(s) + \left(\sum_{i=0}^{m_c} k_i s^i\right) B_p(s) = P(s) = \sum_{i=0}^{n} a_i s^i. \tag{1.72}$$

This is also expressed in matrix form. In order to simplify the expression, the plant is assumed to be 1/3 order and the controller is PID.

$$[l_1 \ l_0 \ k_2 \ k_1 \ k_0]M = [a_4 \ a_3 \ a_2 \ a_1 \ a_0], \tag{1.73}$$

$$M = \begin{bmatrix} d_3 & d_2 & d_1 & d_0 & 0 \\ 0 & d_3 & d_2 & d_1 & d_0 \\ 0 & n_1 & n_0 & 0 & 0 \\ 0 & 0 & n_1 & n_0 & 0 \\ 0 & 0 & 0 & n_1 & n_0 \end{bmatrix},$$

$$l_1 = 1, \quad l_0 = 0.$$

The matrix M is called the Sylvester matrix. It must be clearly understood that matrix M is not necessarily square, and known parameters are on both sides of the equation. If M is square and all left side parameters are unknown, it is equivalent to the pole placement design approach.

The fourth step is to determine the unknown controller parameters in such a way that the coefficients of the characteristic polynomial are in favorable shape. At this stage, the controller and characteristic polynomial are simultaneously determined. This stage is the most important stage in CDM and will be discussed in detail in later chapters after some preparation of mathematics.

The fifth step is to determine the coefficients of reference numerator, m_i, such that the command following characteristics is satisfied. The above five steps are the general design steps in CDM. The fourth step is really the heart of CDM.

1.5 Control Structure

There are many control design approaches, and thus various controllers can be designed even for the same plant and specifications. For such controllers, there always exist equivalent CDM controllers. Thus, if controllers are compared in the common CDM structure, the nature of such control design approaches will become clear. In order to clarify the features of such controllers, the term *control structure* is introduced. The term is defined as the order of such an equivalent CDM controller and the allowable range of controller parameters.

In this section, the control structures of various design approaches will be discussed. **CDM Control Structure** is first explained in order to make the meaning of the control structure clear. Then, **Classical Control** is discussed to clarify its control structure and its limitations. Then, **LQG** approach is discussed to clarify its control structure and compared with that of CDM. Any LQG controller can be expressed by a CDM controller, but the converse is not true. Then **Augmented LQR** is discussed. This structure is not well known, but it produces low-order controllers. Because, for a CDM controller, there always exists an augmented LQR with the same characteristics, it may be called a modern control expression of CDM. Finally, **LQG with Q-parameterization** is discussed. This structure is developed in connection with H_∞ control. It is shown that this structure is also expressed in the CDM standard expression.

CDM Control Structure has already been explained in Sect. 1.4. But some important aspects will be further explained to make the comparison with other approaches easier. The CDM standard block diagram is shown again in Fig. 1.13.

The mathematical model of controller is as follows:

$$G_c(s) = B_c(s)/A_c(s), \quad G_a(s) = B_a(s)/A_c(s), \tag{1.74}$$

$$A_c(s) = l_{n_c} + \cdots + l_1 s + l_0, \tag{1.75}$$

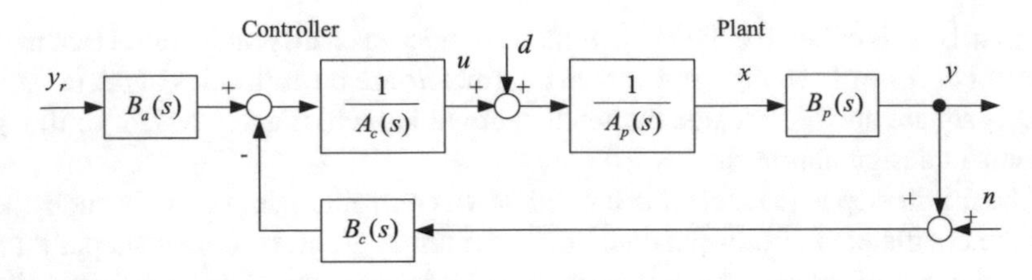

Fig. 1.13 Block diagram of a CDM standard feedback system

$$B_c(s) = k_{m_c} + \cdots + k_1 s + k_0, \tag{1.76}$$

$$B_a(s) = m_{m_a} + \cdots + m_1 s + m_0, \tag{1.77}$$

where $G_c(s)$ is the controller transfer function for feedback signal and $G_a(s)$ is the controller transfer function for reference signal. The controller is m_c/n_c order. In CDM, any order is possible, but usually 2/2 order suffices. Very rarely, 3/3 order or higher are required. The parameters are usually positive, but zero or negative values are possible depending on the situation. Especially, the case for $l_1 = 1, \; l_0 = 0$, such that

$$G_c(s) = \frac{k_2 s^2 + k_1 s + k_0}{l_2 s^2 + s}, \tag{1.78}$$

has the widest application. This controller may be called as a generalized PID. It looks like a PID controller, but it differs in that k_2 and k_1 are not necessarily positive, and that l_2 is not necessarily small, but a value determined by design. By this expansion, control of oscillatory plant becomes much easier. This problem is considered fairly difficult by PID, or even by LQG.

Classical Control has two types of controllers. One is PID controller as follows:

$$G_c(s) = K\left(1 + \frac{1}{T_I s} + T_D s\right), \tag{1.79}$$

where all parameters are positive. PID controller is equivalent to 2/2 order CDM controller with the following restriction.

$$l_2 = 0, \;\; l_1 = 1, \;\; l_0 = 0, \;\; k_2 = K T_D, \;\; k_1 = K, \;\; k_0 = K/T_I. \tag{1.80}$$

By this restriction, it is difficult to make only k_1 equal zero or small negative as often required in the control of the oscillatory plant. In reality, PID control is usually effective for the plant with many time-lags.

The other is the lead-lag compensator as follows:

$$G_c(s) = K\left(\frac{1 + \alpha_1 T_1 s}{1 + T_1 s}\right)\left(\frac{1 + \alpha_2 T_2 s}{1 + T_2 s}\right), \tag{1.81}$$

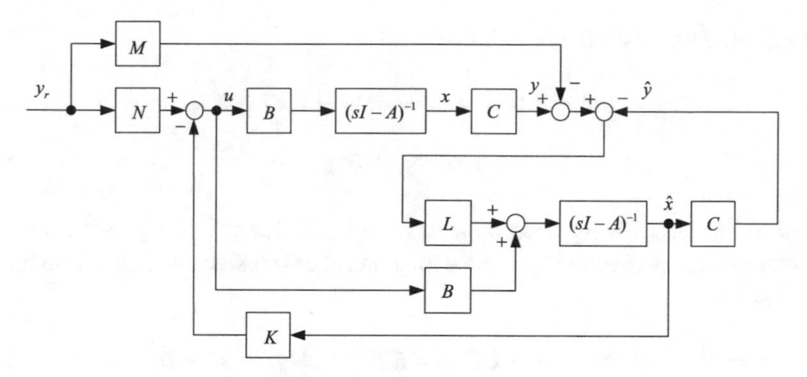

Fig. 1.14 Block diagram of a LQG control

$$\alpha_1 > 1, \;\; 0 < \alpha_2 < 1, \;\; 0 < T_1 < T_2,$$

where the first part is lead compensator and the second part is lag compensator. Lead-lag compensator is equivalent to 2/2 order CDM controller with the following restriction.

$$l_2 = T_1 T_2, \;\; l_1 = T_1 + T_2, l_0 = 1,$$

$$k_2 = K(\alpha_1 \alpha_2 T_1 T_2), \;\; k_1 = K(\alpha_1 T_1 + \alpha_2 T_2), \;\; k_0 = K. \tag{1.82}$$

This is more restrictive than PID in that the roots of denominator/numerator polynomials are real negative, and complex or real positive roots are not allowed. The lead-lag compensator is usually effective for the plant with poles and zeros of negative real, such that the plant is expressed easily by the asymptotic Bode diagram.

In classical control, the order of the controller looks sufficient. However, due to the restriction on parameter range, the power of control is limited. Also, it has to be noticed that the two-degree-of-freedom system has been used in classical control for many years, even before this name is commonly accepted around the 1990s. The controller takes the following form.

$$u = G_a(s) y_r - G_c(s) y, \tag{1.83}$$

where $G_a(s)$ is the controller transfer function for reference signal as explained before.

LQG (Linear Quadratic Gaussian) approach is based on the block diagram of Fig. 1.14. The block diagram can be converted to the CDM standard block diagram of Fig. 1.13.

The plant transfer function is as follows:

$$y = C(sI - A)^{-1} B u = \frac{C \operatorname{adj}(sI - A) B}{\det(sI - A)} u. \tag{1.84}$$

The equation for controller is as follows:

$$(sI - A)\hat{x} = L(y - My_r - C\hat{x}) + Bu, \tag{1.85}$$

$$u = Ny_r - K\hat{x}. \tag{1.86}$$

The u in Eq. (1.85) is replaced by u in Eq. (1.86). Then, $\hat{x}$ is expressed by y and y_r. When this $\hat{x}$ is used in Eq. (1.86), the controller transfer function is obtained as follows:

$$u = Ny_r + K(sI - A + LC + BK)^{-1}[L(My_r - y) - BNy_r]. \tag{1.87}$$

In order to simplify the above equation, the following identity is used.

$$P_o(s) = \det[sI - A + LC] = \det\begin{bmatrix} sI - A + LC + BK & B \\ K & 1 \end{bmatrix} \tag{1.88}$$

$$= \det[sI - A + LC + BK][1 - K(sI - A + LC + BK)^{-1}B].$$

Then, the following result is obtained.

$$u = \frac{P_o(s)Ny_r + K\mathrm{adj}(sI - A + BK + LC)L(My_r - y)}{\det(sI - A + LC + BK)}. \tag{1.89}$$

Then, the plant/controller polynomials are obtained as follows:

$$A_p(s) = \det(sI - A), \tag{1.90}$$

$$B_p(s) = C\mathrm{adj}(sI - A)B, \tag{1.91}$$

$$A_c(s) = \det(sI - A + BK + LC), \tag{1.92}$$

$$B_c(s) = K\mathrm{adj}(sI - A + BK + LC)L, \tag{1.93}$$

$$B_a(s) = P_o(s)N + B_c(s)M. \tag{1.94}$$

The related characteristic polynomials and Diophantine equation are given as follows:

$$P_c(s) = \det(sI - A + BK), \tag{1.95}$$

$$P_o(s) = \det(sI - A + LC), \tag{1.96}$$

$$P(s) = P_o(s)P_c(s) = A_c(s)A_p(s) + B_c(s)B_p(s). \tag{1.97}$$

In design, state feedback characteristic polynomial $P_c(s)$ and observer characteristic polynomial $P_o(s)$ are first specified. Then, K and L can be calculated by various methods such as Ackerman's formula. This leads to direct calculation of $A_c(s)$ and $B_c(s)$ by Eqs. (1.92) and (1.93). But much easier way is to solve the Diophantine equation Eq. (1.97) for $A_c(s)$ and $B_c(s)$.

The first shortcoming of the LQG control structure is that the controller order is fixed as $(n_p - 1)/n_p$ order. The low-order controllers are excluded from the outset. The second shortcoming is that the controller parameters can be positive or negative depending on the situation. If the nature of $B_p(s)$ is not properly reflected in the choice of $P_c(s)$, the designed controller polynomials have negative parameters. Then, robustness is greatly deteriorated. There is no guarantee of robustness in LQG. Because of these two shortcomings, LQG is not considered as a good design practice, contrary to the common understanding. Because $B_a(s)$ can be adjusted by M and N, LQG is a two-degree-of-freedom system.

Augmented LQR (Augmented Linear Quadratic Regulator) approach is an improvement of LQG. In order to simplify notations, a 2/3 order plant is chosen, where plant polynomials are given as follows:

$$A_p(s) = s^3 + d_2 s^2 + d_1 s + d_0, \tag{1.98}$$

$$B_p(s) = n_2 s^2 + n_1 s + n_0. \tag{1.99}$$

By augmenting the system by order 2, the augmented plant is given as follows:

$$s^2 u = v, \tag{1.100}$$

$$A_p(s) x = u, \tag{1.101}$$

$$y = B_p(s) x, \tag{1.102}$$

where v is the new input to the augmented plant. This equation is rewritten in the following form by the introduction of a new variable $w = B_p(s)u$.

$$s^2 u = v, \tag{1.103}$$

$$w = B_p(s)u = (n_2 s^2 + n_1 s + n_0)u, \tag{1.104}$$

$$A_p(s)y = (s^3 + d_2 s^2 + d_1 s + d_0)y = w. \tag{1.105}$$

In these equations of augmented plant, we define $u_1 = su,\ u_0 = u,\ y_2 = s^2 y,\ y_1 = sy,\ y_0 = y$, and the state vector $\mathrm{x} := [u_1\ u_0\ y_2\ y_1\ y_0]^T$. The state equation becomes as follows:

$$\dot{\mathrm{x}} = A\,\mathrm{x} + Bv, \qquad y = C\,\mathrm{x}, \tag{1.106}$$

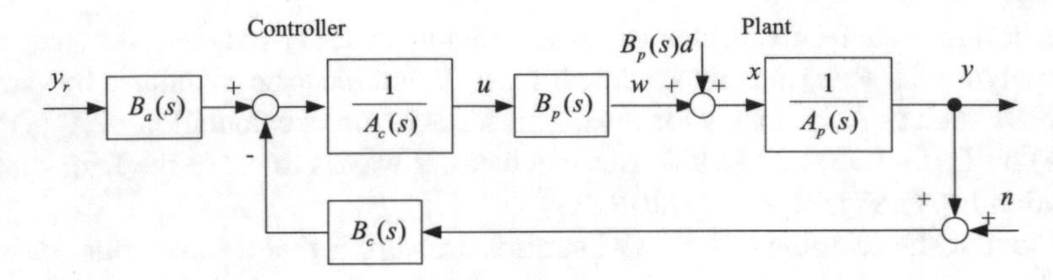

Fig. 1.15 Block diagram of an Augmented LQR control

$$A = \begin{bmatrix} 0 & 0 & 0 & 0 & 0 \\ 1 & 0 & 0 & 0 & 0 \\ n_1 & n_0 & -d_2 & -d_1 & -d_0 \\ 0 & 0 & 1 & 0 & 0 \\ 0 & 0 & 0 & 1 & 0 \end{bmatrix}, \quad B = \begin{bmatrix} 1 \\ 0 \\ n_2 \\ 0 \\ 0 \end{bmatrix},$$

$$C = [0 \;\; 0 \;\; 0 \;\; 0 \;\; 1].$$

Then, the following state feedback control is considered.

$$v = (m_2 s^2 + m_1 s + m_0) y_r - (l_1 u_1 + l_0 u_0) - (k_2 y_2 + k_1 y_1 + k_0 y_0). \tag{1.107}$$

Considering $v = s^2 u$, the controller is exactly the same as the CDM controller as follows:

$$A_c(s) u = B_a(s) y_r - B_c(s) y, \tag{1.108}$$

$$A_c(s) = s^2 + l_1 s + l_0,$$

$$B_c(s) = k_2 s^2 + k_1 s + k_0,$$

$$B_a(s) = m_2 s^2 + m_1 s + m_0.$$

The block diagram for this system is shown in Fig. 1.15 with proper addition of disturbance d and noise n. The comparison with Fig. 1.13 reveals that this is equivalent to the CDM standard block diagram. Thus, the augmented LQR approach retains exactly the same control structure as CDM. When a CDM controller is designed, it is always possible to find an equivalent augmented LQR controller. Such augmented LQR is of a lower order, and robustness is guaranteed because it is CDM design itself. But it is very difficult to design a CDM controller from the augmented LQR controller simply because the present LQR design theory does not provide reliable weight selection rules.

LQG with Q-Parameterization is a variation of LQG controller. This structure has been developed in the effort of finding all possible stabilizing controllers at H_∞ controller design. The block diagram is shown in Fig. 1.16, where a path from η to r is added to the standard LQG. The Q parameter is a rational function of s,

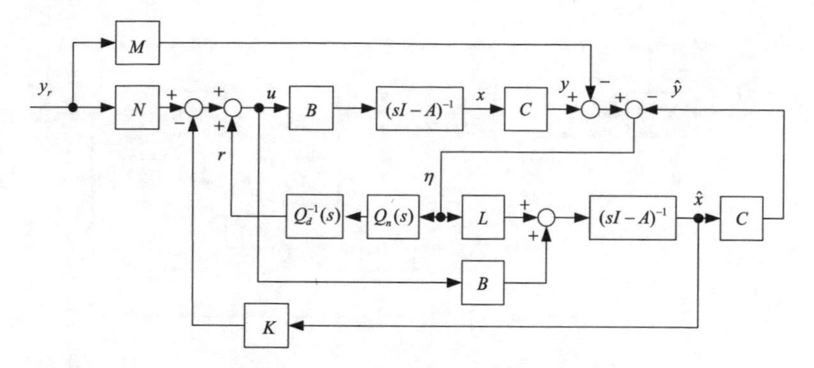

Fig. 1.16 Block diagram of a LQG control with Q-parameterization

and it is expressed as the ratio of numerator polynomial $Q_n(s)$ and denominator polynomial $Q_d(s)$. The block diagram is expressed in polynomial form as Fig. 1.17 by the following process. The related equations are as follows:

$$\eta = y - My_r - C\hat{x}, \tag{1.109}$$

$$(sI - A)\hat{x} = L\eta + Bu, \tag{1.110}$$

$$u = Ny_r - K\hat{x} + r, \tag{1.111}$$

$$r = Q_d^{-1}(s)Q_n(s)\eta. \tag{1.112}$$

Eliminating $\hat{x}$ in Eqs. (1.109) by (1.110) yields

$$[1 + C(sI - A)^{-1}L]\eta = y - My_r - C(sI - A)^{-1}Bu. \tag{1.113}$$

By matrix-to-determinant conversion similar to Eq. (1.88),

$$P_o(s)\eta = A_p(s)(y - My_r) - B_p(s)u. \tag{1.114}$$

Substituting Eqs. (1.109) and (1.111) into Eq. (1.110), we have

$$(sI - A + LC + BK)\hat{x} = L(y - My_r) + B(Ny_r + r). \tag{1.115}$$

Eliminating $\hat{x}$ in Eqs (1.111) by (1.115) gives

$$\begin{aligned} u = {} & [1 - K(sI - A + LC + BK)^{-1}B](Ny_r + r) \\ & + K(sI - A + LC + BK)^{-1}L(My_r - y). \end{aligned} \tag{1.116}$$

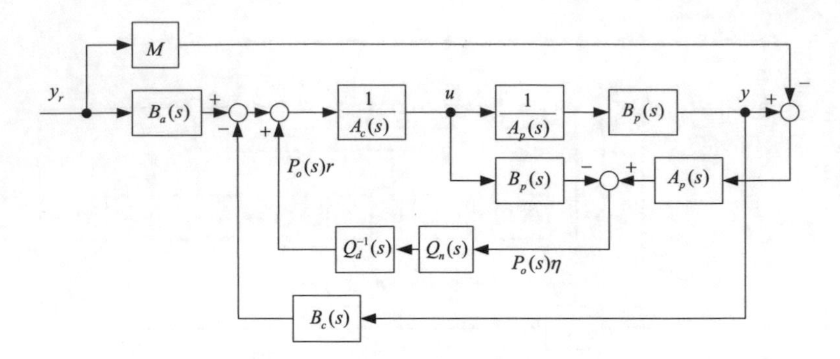

Fig. 1.17 Polynomial block diagram for LQG with Q-parameterization

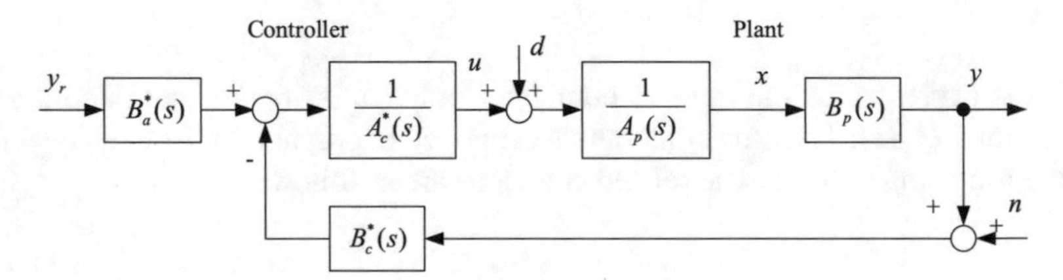

Fig. 1.18 CDM standard block diagram for LQG with Q-parameterization

By matrix-to-determinant conversion similar to Eq. (1.88),

$$A_c(s)u = B_a(s)y_r - B_c(s)y + P_o(s)r. \tag{1.117}$$

Equations (1.114) and (1.117) result in another representation of LQG with Q-parameterization, as shown in Fig. 1.17.

Then, Fig. 1.17 is converted to the CDM standard block diagram as in Fig. 1.18. The new controller polynomials $A_c^*(s)$, $B_c^*(s)$, and $B_a^*(s)$ are expressed as follows:

$$A_c^*(s) = Q_d(s)A_c(s) + Q_n(s)B_p(s), \tag{1.118}$$

$$B_c^*(s) = Q_d(s)B_c(s) - Q_n(s)A_p(s), \tag{1.119}$$

$$B_a^*(s) = MB_c^*(s) + NQ_d(s)P_o(s). \tag{1.120}$$

When $Q_d(s) = 1$ and $Q_n(s) = 0$, the controller is the LQG controller expressed in CDM standard expression. The new characteristic polynomial $P^*(s)$ is given as follows:

$$\begin{aligned} P^*(s) &= A_c^*(s)A_p(s) + B_c^*(s)B_p(s) \\ &= Q_d(s)[A_c(s)A_p(s) + B_c(s)B_p(s)] = Q_d(s)P(s). \end{aligned} \tag{1.121}$$

Thus, $P^*(s)$ retains all the poles of LOG together with the added poles due to $Q_d(s)$. Then, the stability is guaranteed. The new sensitivity function $S^*(s)$ and complementary sensitivity function $T^*(s)$ are expressed as follows:

$$S^*(s) = S(s) + \delta S(s), \quad T^*(s) = T(s) - \delta S(s), \tag{1.122}$$

where

$$\delta S(s) = \frac{Q_n(s)B_p(s)A_p(s)}{Q_d(s)P(s)}. \tag{1.123}$$

This shows that the function of Q parameter is to modify sensitivity and complementary sensitivity functions without changing stability. The Q parameter is to be used for the small modification to the standard LQG design.

When a large modification is required such that poles have to be placed to new locations, it is still possible by assigning $Q_d(s)$ and $Q_n(s)$ as follows:

$$Q_d(s) = P_1(s) = A_{c1}(s)A_p(s) + B_{c1}(s)B_p(s), \tag{1.124}$$
$$Q_n(s) = A_{c1}(s)B_c(s) - B_{c1}(s)A_c(s), \tag{1.125}$$

where $A_{c1}(s)$ and $B_{c1}(s)$ are denominator and numerator polynomials of desired controller, and $P_1(s)$ is the desired characteristics polynomial. In this situation, the controller polynomials and the characteristic polynomial are given as follows:

$$A_c^*(s) = A_{c1}(s)P(s), \tag{1.126}$$
$$B_c^*(s) = B_{c1}(s)P(s), \tag{1.127}$$
$$P^*(s) = P_1(s)P(s). \tag{1.128}$$

By cancellation of $P(s)$, the system behaves as the desired system. However, such modification is not recommended because of unnecessary complication. It is much simpler to design a new controller.

In conclusion, LQG with Q-parameterization is to be used for the modification of sensitivity and complementary sensitivity functions, while the characteristic polynomial is retained as it is with the addition of the poles of $Q_d(s)$. Large modification of characteristic polynomial is not recommended. At any event, LQG with Q-parameterization is expressed in the CDM standard expression.

1.6 Historical Background

The CDM has been developed on many previous ideas and experiences in control system design. Briefly stated, however, CDM may be regarded as a practical extension of the idea of J. C. Maxwell *On Governors* [9]. Some of the important topics will be covered in **Brief History**. In CDM, three new features were added to the past experiences. **Three New Features** is the detailed history of how such new features were developed. **Present and Future Trend of CDM** will briefly discuss the present activity and future trend.

Brief History related to CDM will be explained hereafter. The first treatment of the polynomial approach is *On Governors* by J. C. Maxwell in 1868 and the Routh stability criterion in 1877 [2], where the stability is analyzed using the coefficients of the characteristic polynomial. However, the original form of stability criterion has been kept in the original form, and no further conspicuous efforts have been made to make this approach a workable design methodology until Lipatov's and Sokolov's work [10].

In the 1950s, the frequency response method was widely used in control system design. During that period, it was commonly recognized that, for good system design, such criteria as phase margin or gain margin were not sufficient and frequency characteristics of the open-loop transfer function should have proper shape for a fairly wide frequency range [11].

Chestnut pointed out, in his celebrated book [12], the importance of the relative location of break points and the change of slope at the break points of the asymptotic Bode diagram (gain plot only) for the open-loop transfer function. He proposed a design method based on these findings. His proposal was very practical and has been widely used in the industry not only in the 1950s but even today.

The rule of thumb, such that the asymptotic gain plot should intersect the 0 dB line at the slope of -20 dB/decade, the change of slope at each break points should be ± 20 dB/decade, and the break points should be separated at least by a factor of two, has been widely used in the practical design of simple control systems.

For such simple control systems, the separation of break points approximately corresponds to the stability index. The rule of thumb that the break points should be separated at least by a factor of two roughly corresponds to specifying stability index γ_i to be larger than two. The effort to make this rule applicable to more complex systems has led later to the adoption of stability index rather than the break points, and the adoption of the coefficient diagram rather than the Bode diagram, and finally, culminated in CDM.

Graham [13] made intensive research to find the relation between the coefficients of characteristic polynomials and the transient responses and proposed standard forms for desirable characteristic polynomials. This is commonly called ITAE (Integral Time Absolute Error) standard form. The values of coefficients of this standard form are similar, but a little more oscillatory, compared to the proposed values for CDM.

The shortcoming of ITAE as a design approach is due to its lack of flexibility. Because it gives a standard form for each order of the characteristic polynomial, it is very inconvenient when the order varies in the course of design. Because it gives only one standard form and fails to show the way to modify it when necessary, an unnecessarily unrobust controller can be designed at certain occasion [2, see Example 7.21 in 4th Ed.]. Even with these shortcomings, this work is a great contribution to the practical control design.

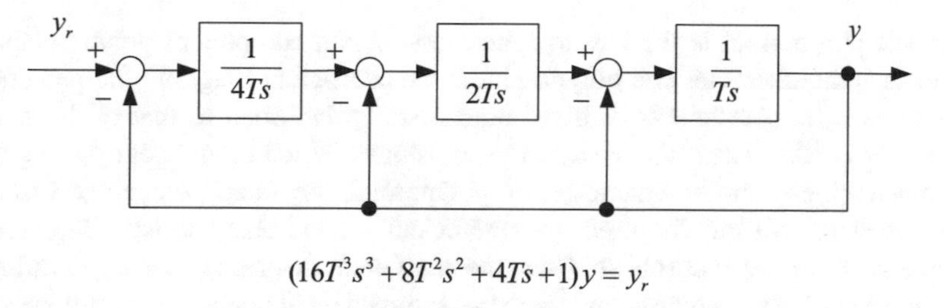

Fig. 1.19 Kessler's canonical multi-loop structure

Around this time, Kessler made intensive efforts to establish synthesis procedures for a multi-loop control system and came out with a standard form, commonly called *Kessler's canonical multi-loop structure* [5], as shown in Fig. 1.19. The proposed system is more stable and robust compared with ITAE standard form and is more fitted to the design of multi-loop control systems. For this reason, it has been widely used in steel mill control. However, Kessler's standard form amounts to proposing all stability index γ_i to be 2 and has an unnecessary overshoot of 8%. It was later found that the no-overshoot condition can be easily obtained by a small modification of making γ_1 larger to $\gamma_1 = 2.5$. The CDM incorporates this modification.

Various researchers have developed the similar idea in Europe [6, 7, 14]. The stability index was called by different names; *damping factor* by Kessler, *double ratio* by Brandenburg, and *characteristic ratio* by Naslin. Kitamori proposed an improved version of the approach of Graham, where the specification of the characteristic polynomial was given for low-order coefficients [15]. In this way, the flexibility of the design was greatly improved.

The stability of control systems can be analyzed by the Routh or Hurwitz criterion utilizing coefficients of characteristic polynomials. However, in this way, the effect of the variation of coefficients on stability is not clearly seen. Lipatov and Sokolov proposed sufficient conditions for stability and instability [10]. Because of its simplicity, the relation of stability and instability with respect to the coefficients of the characteristic polynomials becomes very clear. These conditions are integrated into the design procedures of CDM.

Especially, they become powerful design tools when it becomes clear that these conditions can be easily interpreted on the coefficient diagram graphically. At the time when the coefficient diagram was introduced, it was already known that the curvature is closely related to stability indices γ_i and thus to stability. But the graphical interpretation of Lipatov's stability condition gives much more accurate information about stability, and more sophisticated design becomes possible. The inclination of the coefficient diagram is closely related to the equivalent time constant τ, and thus response speed. Also, graphical procedures to estimate the time responses and frequency responses are later developed with the accumulation of experience. In the construction of the coefficient diagram, designers have to compute the *component polynomials*, which constitute the characteristic polynomial. Then, it becomes evident that the comparative magnitude of such component polynomials to the characteristic polynomial is the key to robustness. When component polynomials are lower than the characteristic polynomial at the coefficient diagram, the percentage variation of the coefficients of the characteristic polynomial to that of the related parameter is low. Then, the system is very robust. When component polynomials are much upper than the characteristic polynomial, the small percentage variation of parameters will greatly affect the coefficients of the characteristic polynomial. Such a system is very unrobust. Thus, the coefficient diagram has enough information about stability, response, and robustness; three key elements of control systems. With experience, designers can visualize the total picture of the control system at a glance of the coefficient diagram. The coefficient diagram is the most important ingredient in CDM. It plays the same role as the Bode diagram in the classical control theory.

In control system design, classical control theory and modern control theory are used. In classical control, transfer functions are used for system representation and the design proceeds usually by the Bode diagram or root locus plot. In modern control, the state-space representation is used and design proceeds by solving the Riccati equation. But there is a third approach called the algebraic approach, or sometimes called the polynomial approach. In this approach, polynomials are used for system representation, and the design is based on the solution of the Diophantine equation. The advantage of this approach is that polynomials are easier to understand compared with state space. The Diophantine equation is linear and easy to solve, while the Riccati equation is quadratic. Although various researchers have made intensive studies so far, these studies mainly handle the problems after the characteristic polynomials have been specified. No serious attempt has been made to select the characteristic polynomial best fitted to the given problem; characteristic polynomials are chosen by pole assignment more or less in an ad hoc manner. In CDM, the utmost concern is to find the proper characteristic polynomial best fitted to the given problem. After such characteristic polynomial is found, this algebraic approach greatly simplifies the process of finding the specific controller, and this approach is adopted in CDM.

Three New Features are added in the development of CDM. Many ideas and past experiences were adopted in CDM as already explained in the brief history. However, in order to clarify the features of CDM, three new features added in the course of CDM development will be explained hereafter.

The first addition is the introduction of the coefficient diagram, where the three important features of the control system, namely stability, response, and robustness, are represented by a single diagram in a graphical manner. Thus, the understanding of the total system becomes much easier. The second addition is the improvement of Kessler's standard form, by which the 8% overshoot in Kessler becomes no-overshoot. The third addition is the introduction of Lipatov's sufficient condition for stability in the form compatible with CDM.

In order to clarify the background of such features, the author (Shunji Manabe) wishes the readers will allow him to explain some of his personal experiences. His first encounter with automatic control was at the Ohio State University during 1952–1954. In the lecture by Prof. Warren, he met the Routh criterion and marveled about its beauty. Also, he met an Air Force captain, who was working on a Master's thesis.

His work was to meticulously calculate and draw time response curves for third-order system. It was quite a work when the slide rule was the only means of computation. The author was later surprised that the same graphs were in Graham's paper.

After coming back to Japan, he started to work in the field of motor speed control at Mitsubishi Electric Research Laboratory. At first, he tried the Nyquist criterion he learned in USA, but it was very inconvenient due to the computational burden. Then, he met Chestnut's book, where the Bode diagram was used and the importance of the separation of break points was stressed. Around that time, a senior engineer in the control field told him that the key to stability was to separate the time constants in the controller and plant at least by a factor of two. This rule of thumb matched with Chestnut's findings and also with the results of analog computer simulation.

Toward the end of the 1950s, the author was involved in the 45,000 HP transonic wind tunnel project. In order to validate the feasibility of the control system, a 1/100 scale model test was performed. In an effort to find the plant transfer function, a vibration mode of about 20 Hz was observed. Because the model was small, vibration is at high frequency and no harmful interaction with the control loop was observed. However in the real plant, the vibration frequency would become lower, and the interaction with the control loop was anticipated. A careful study was made using the Bode diagram and analog computer simulation. However, the shape of the Bode diagram was quite different from the one for ordinary control systems. The transfer function of the plant $G_p(s)$ was in the following form.

$$G_p(s) = \frac{(s/\omega_1)^2 + 1}{s[(s/\omega_2)^2 + 1]}, \tag{1.129}$$

where ω_1 and ω_2 are anti-resonance and resonance frequency, respectively. This system is generally called as two-mass resonant system.

Around that time, Taro Yoshida of Mitsubishi Electric informed him through private communication "Better control can be realized by splitting the feedback signal into two parts and placing them before and after saturation separately." The author had just read Tustin's paper on position control of the massive object. In such motion control, the saturation of the actuator is predominant. At saturation, the gain is much reduced. For this reason, the phase margin has to be maintained not only around the cross-over frequency, but also far below it. Tustin proposed an open-loop transfer function $G(s)$ as follows:

$$G(s) = A_0(\omega_c/s)^k, \quad 1 < k < 2, \tag{1.130}$$

where $A_0 = 1$ without saturation, and A_0 is much reduced with saturation. Such system is later, in the 1990s, named as *Fractional-Order System*. The author saw immediately that the proposal by Yoshida was a practical implementation of the fractional-order system. He made an intensive study of this system and came out with the result that the best value of k is around 1.4, or -28 dB/decade slope of asymptotic Bode diagram [16]. This was a very simple and practical design criterion and had been used successfully in the design of tracking radar antenna control and other motion control. Around these periods, design by asymptotic Bode diagram of the open-loop transfer function is the main design technology. In such problems, the plants were usually characterized by real poles and zeros.

Toward the end of the 1970s, the author was involved in the spacecraft attitude control. There were two vibration problems. One was the interaction of flexible appendage structure like a solar paddle to the spacecraft attitude. The transfer function of the plant is similar to the two-mass resonant system as Eq. (1.129). These problems could be solved much in the same manner as the control of a two-mass resonant system. The other was the nutation attenuation control of spacecraft with momentum wheel. The transfer function of the plant $G_p(s)$ is in the following form.

$$G_p(s) = \frac{1}{s^2 + 1}, \tag{1.131}$$

where the nutation frequency is normalized as 1 rad/s, and the input is the momentum of the second small wheel. The output is the roll angle of the spacecraft. Generally speaking, the best policy for vibration control is to weaken the spring force and add damping force. The control effort to weaken the spring amounts to the introduction of positive feedback or the non-minimum -phase controller. The controllers designed at that time were non-minimum-phase controllers [17]. The design method was pole assignment approach. Then, robustness was to be analyzed separately.

The controller the author chose is in the following form.

$$G_c(s) = \frac{k_2 s^2 + k_1 s + k_0}{l_2 s^2 + s}. \tag{1.132}$$

This controller looks like a PID, but it differs in that k_1 is zero or weakly minus and l_2 is fairly large. This type of controller was named as *Zero-PID Controller* [18].

During the 1980s, the author was looking for a design approaches, whereby controllers for three different control problems, namely, two-mass resonant system Eq. (1.129), fractional-order system Eq. (1.130), and nutation control Eq. (1.131), can be designed in a unified manner. This led to the idea of modifying classical control from open-loop to closed-loop [19] without much success. In July 1985, the author visited Cambridge University and was able to study the work of J. C. Maxwell with the kind help of Prof. MacFahlen. The author thought that the extension of the approach by Maxwell might be the answer to control design problems. He did not have much confidence in the modern control approach, because of the difficulty in designing PID equivalent controllers with proper weight selection.

On some late afternoon in May 1988, he was walking to his office building. Suddenly, the image of the coefficient diagram came up to his mind. He suddenly realized the controllers could be designed in a unified manner for three different plants. The first coefficient diagram drawn on his notebook was dated May 13, 1988. The above is his long journey to the coefficient diagram.

The author started to teach control theory at Tokai University from 1990. He was teaching the Kessler standard form as the design basis. In the summer of 1990, a

Master's student suggested to him that the no-overshoot condition can be achieved by increasing γ_1. The student discovered this important result in his effort to complete a control system design assignment. After careful study, it was found that the no-overshoot condition of γ_1 is 2.7 for the third-order system and 2.5 for the fourth-order system. The standard value of γ_1 is chosen to be 2.5 for the sake of simplicity and convenience in handling the number; such that the reciprocal of γ_1 is $1/2.5 = 0.4$. Coefficient diagram method was first introduced in a symposium with $\gamma_1 = 2.7$ [20] and later published in a journal with $\gamma_1 = 2.5$ [21]. Further improvement has been made on the selection of stability index, and a more flexible standard is in use today, where the robustness issue is fully incorporated.

Incidentally, the author heard about Kessler's work since the late 1970s, but it was September 18, 1992, when he was able to locate Kessler's paper at the VDI office (Association of German Engineers) of Düsseldorf.

Around 1989, the author happened to read a paper by Bose [22] and learned about the work by Lipatov and Sokolov [10] on sufficient condition for stability and instability. He immediately saw the importance of the theorem as the mathematical basis of CDM. In order to make the graphical interpretation of stability condition, he introduced *Stability Limit*. The stability could be checked by the comparison of stability index γ_i to stability limit γ_i^*. However, this relation was not convenient, because computations of γ_i and γ_i^* were necessary. After some years, this condition was converted to the relation of coefficients with graphical interpretation [23]. By this interpretation, truly coefficient diagram-based design became possible.

Present and Future Trend of CDM will be briefly presented hereafter. With the maturity of CDM from crude infancy [21] to a sound control design theory, many successful designs were reported. The author planned to build something unusual to show the effectiveness of CDM. He built a single sensor (only angle measurement) inverted pendulum of 20 cm length on a toy model car. The controller was a second-order unstable controller composed of 4 power transistors and 12 operational amplifiers packed in 3 IC packages. The car was demonstrated at a conference, and the effectiveness of CDM in designing such an unusual controller was reported [24]. Other design examples and interpretation of past successful designs in CDM were also reported [24–26]. Tanaka [27] developed an independently similar approach by specifying α parameter, which is reciprocal of stability index, with a successful application to a gas turbine controller design [28]. Hori [29] used the stability index for the controller design of the two-mass resonant system. The relation of CDM to LQG was clarified, and CDM-equivalent augmented LQR was derived [23]. The paper by Lipatov was further studied and analyzed [30]. By these efforts, the mathematical foundation of CDM became clearer. These results were summarized, and the design procedures of CDM were presented in a consistent manner in the IFAC symposium [23].

The first organized session for CDM was organized by Y. C. Kim at Korea Automatic Control Conference in October 1998. Similar organized sessions were held in the 3rd and 4th Asian Control Conference in 2000 and 2002 [31, 32]. Various authors contributed their experiences with CDM. The application of CDM to MIMO is also in progress. During this effort, it was found that CDM and polynomial design approach have much in common. The cross-fertilization of the two approaches will be much beneficial to both sides. CDM will be much rewarded with the rich results in the polynomial design of the last few decades.

1.7 Summary

The important points in this chapter are summarized in the following. **Basic Philosophy** discusses three basic philosophies of CDM. At first, "Control is compromise" is explained. Then, it is emphasized that "Feedback control is only a small part of control", and too much sophistications are not justified. Finally, the algebraic approach is briefly explained. It is the third approach after classical control and modern control. The CDM is one of the algebraic approaches, but it is unique in adopting the simultaneous design approach, while classical control adopts the outward approach, and modern control adopts the inward approach.

Simple Design Problem illustrates a simple design example by CDM of a typical position control system. The definitions of stability index and equivalent time constant are given. The controller is designed systematically. The coefficient diagram is introduced. It is also shown that the step responses are closely related to the shape of the coefficient diagram. Finally, the coefficient-shaping approach is explained. With this approach, the controller and the closed-loop characteristics, such as step response, are designed simultaneously with the help of the coefficient diagram.

System Representation explains that polynomial expression used in CDM is superior to transfer function expression or state-space expression in many respects. The polynomial expression is nothing but the differential equation, and its extension to the nonlinear system or to the fractional-order system is not difficult. The polynomial expression is as easy to understand as the transfer function expression, but there is no ambiguity. The polynomial expression is as accurate as the state-space expression, but the clumsiness in handling is avoided.

Outline of Design Process briefly explains the outline of the CDM design process. At first, the mathematical model for CDM design is shown. Then, the four input–output relations, on which CDM design is based, are explained. After that, the pole-zero cancellation problem is discussed, and it is shown that undesirable pole-zero cancellation is prevented beforehand in CDM. Finally, the actual design step is shown. However, the simultaneous design of the controller and characteristic polynomial, the most important part of CDM design, is left for later chapters, because it can be explained effectively only through examples.

Control Structure explains that various types of controllers designed by different design approaches can be transformed to equivalent CDM standard controllers. The control structure is defined as the order and the range of controller parameters of such equivalent CDM controllers. At first, the control structure of the CDM controller is explained, and it is shown that the parameter range is very wide. Then, the control structure of classical control is explained, and it is shown that the parameter range is much limited. Then, the control structure of the LQG design is discussed, and it is shown that the low-order controller cannot be designed and that there is no guarantee of robustness. After that, the control structure of augmented LQR (Linear quadratic regulator) is discussed, and it is explained that such controller has a one-to-one correspondence with the CDM controller. Finally, the control structure of LQG with Q-parameterization is discussed, and the equivalent CDM controller is derived.

Historical Background shows a short history of the development of CDM. At first, a brief history of control theories, which constitute the background of CDM, is shown. Then, the experiences of the author, which result in three new additions; namely the coefficient diagram, the revision of the Kessler standard form, and the graphical interpretation of the sufficient condition of stability by Lipatov, are explained in more detail. Finally, the present and future trends of CDM are briefly discussed.

At the end of this chapter, the author wishes to answer the question of the reader, "What is CDM?" The CDM is one of the algebraic approaches over a polynomial ring, where the coefficient diagram is used instead of the Bode diagram, and the sufficient condition for stability by Lipatov is used instead of the Routh stability criterion. CDM has six features as follows:

1. CDM is one of the algebraic approaches over a polynomial ring. It is the third control design approach after classical control and modern control.
2. Polynomials or polynomial matrices are used in system representation. The polynomial expression is easy to understand as a transfer function in classical control and as accurate as state-space expression in modern control.
3. The controller and the characteristic polynomial are simultaneously designed, whereby robustness is incorporated to stability and response. Classical control uses the outward approach, where the controller is first assumed and the closed-loop characteristics are later confirmed. Modern control uses the inward approach, where the closed-loop characteristics are first given and the controller is determined accordingly. Different from these approaches, CDM uses the simultaneous design approach.
4. The coefficient diagram is fully utilized in CDM. The coefficient diagram contains sufficient information about stability, response, and robustness in graphical form. Thus, the graphical design approach is possible. The coefficient diagram plays the same role in CDM as the Bode diagram in classical control.
5. The sufficient condition for stability by Lipatov is used as the theoretical basis. The condition is expressed graphically in the coefficient diagram. This condition is used in CDM instead of the Routh criterion.
6. The Kessler standard form is further modified and used as the standard form of CDM. The main modification is to make $\gamma_1 = 2.5$, but further modification is made so that the robustness issue is incorporated in the design.

References

1. Chen CT (1987) Introduction to the linear algebraic method for control system design. IEEE Contr Syst Mag 7(5):36–42
2. Franklin CF, Powell JD, Emami-Naeini A (2015) Feedback control of dynamic systems, 7th edn. Pearson Education Ltd., Essex, England
3. Kailath T (1980) Linear systems, pp 306–310. Prentice-Hall, Englewood Cliffs, NJ
4. Kucera V (1979) Discrete linear control: the polynomial equation approach. Wiley, Hoboken
5. Kessler C (1960) Ein Beitrag zur Theorie mehrschleifiger Regelungen. Regelungstechnik 8(8):261–266
6. Naslin P (1969) Essentials of optimal control. Illife Books Ltd., London
7. Brandenburg G, Brueckel S (1960) The damping optimum and analytical pole assignment approach using double ratios. Unpublished, refer to
8. Franklin CF, Powell JD, Emami-Naeini A (1994) Feedback control of dynamic systems, 3rd edn. Addison-Wesley, Boston
9. Maxwell JC (1868) On governors. Proc R Soc London 16:270–283
10. Lipatov AV, Sokolov NI (1978) Some sufficient conditions for stability and instability of continuous linear stationary systems. Translated from Automatika i Telemekhanika 9:30–37, 1978; Autom. Remote Contr (1979) 39:1285–1291
11. Tustin A et al (1958) Position control of massive objects. Proc IEEE: Part C 105(1):1–57
12. Chestnut H, Mayer RW (1951) Servomechanism and regulating system design, vol 1, Chap 14. Wiley, New York
13. Graham D, Lathrop RC (1953) The synthesis of optimum transient response: criteria and standard forms. AIEE Trans: pt II 72:273–288
14. Zaeh M, Brandenburg G (1987) Das erweiterte Daempfungsoptimum. Automatisierungstechnikat 35(7):257–283
15. Kitamori T (1979) A method of control system design based upon partial knowledge about controlled process. Trans SICE 15(4):549–555
16. Manabe S (1960) The non-integer integral and its application to control systems. J Inst Electr Eng Jpn 80(860):589–597 (Japanese); ETJ of Japan (1961) 6(3):83–87, (Brief English version)
17. Terasaki RM (1967) Dual reaction wheel control of spacecraft pointing. Symposium on attitude stabilization and control of dual-spin spacecraft. SAMSO and Aerospace Corporation, El Segundo, CA, pp 185–196
18. Manabe S, Tsuchiya T, Inoue M (1981) Zero PID control for bias momentum satellites. In: Proceedings of the 8th IFAC world congress, Kyoto, Japan, August 24–28, vol 14, issue 2, pp 2217–2223 (1981)
19. Manabe S (1983) Control theory from the user side. Japan Society of Mechanical Engineers: Lecture Course, no. 560, June 23–24, pp 31–45 (1983)
20. Manabe S (1990) An unified interpretation of classical, optimal, and H-infinity control. In: Proceedings of the 7th SICE symposium on guidance and control in aerospace, Tokyo, Japan, Nov 28–29, pp Toku1–10 (1990)
21. Manabe S (1991) An unified interpretation of classical, optimal, and H-infinity control. J SICE 30(10):941–946
22. Bose NK, Jury EI, Zaheb E (1988) On robust Hurwitz and Schur polynomials. IEEE Trans Autom Control 33(12):1166–1168
23. Manabe S (1998) The coefficient diagram method. In: Proceedings of the 14th IFAC symposium on automatic control in aerospace, Seoul, Korea, Aug 24–28, pp 199–210 (1998)
24. Manabes S (1994) A low-cost inverted pendulum system for control education. In: Proceedings of the 3rd IFAC symposium on advances in control education, Tokyo, Japan, August 1–2, pp 21–24 (1994)
25. Manabe S (1997) The application of coefficient diagram method to the ACC benchmark problem. In: Proceedings of the 2nd Asian control conference, Seoul Korea, July 22–25, pp II135–138 (1997)

26. Manabe S (1998) Analytical weight selection for LQ design. In: Proceedings of the 8th Workshop on Astrodynamics and Flight Mechanics, Sagamihara, ISAS, July 23–24, pp 237–246 (1998)
27. Tanaka Y (1992) α parameter for robust control design. Trans SICE 28(12):1501–1503
28. Tanaka Y, Ashikaga M (1992) A low-sensitive robust a gas turbine. Trans SICE 28(2):255–263
29. Hori Y (1994) 2-mass system control based on resonance ratio control and Manabe polynomials. In: Proceedings of the first Asian control conference, Tokyo, Japan, 27–30 July, no 3, pp 741–744 (1994)
30. Manabe S (1999). Sufficient condition for stability and instability by Lipatov and its application to the coefficient diagram method. In: Proceedings of the 9th workshop on astrodynamics and flight mechanics, Sagamihara, ISAS, July 22–23, pp 440–449 (1999)
31. Manabe S, Kim YC (2000) Recent development of coefficient diagram method. In: Proceedings of the 3rd ASCC, Shanghai, China, July 3–7, pp. 2055–2060 (2000)
32. Manabe S (2002b) Brief tutorial and survey of coefficient diagram method. In: Proceedings of the 4th Asian control conference, Singapore, September 25–27, pp 1161–1166 (2002)

Chapter 2
Basics of Coefficient Diagram Method

Abstract In this chapter, the fundamental properties which are necessary for CDM design will be presented. In **Mathematical Relations**, definitions of symbols and their basic relations are briefly discussed. In **Coefficient Diagram**, the way to draw coefficient diagrams is first introduced and then the interpretation of the coefficient diagram is presented. In **Stability Condition**, the sufficient conditions for stability and for instability by Lipatov are presented. They are the simplification of the stability condition by Routh and constitute the mathematical basis of CDM. In **Canonical Transfer Function**, some special transfer functions, which have special relations with characteristic polynomials, are defined as canonical transfer functions. By use of such functions, the characteristics of general transfer functions can be clarified. In **Standard Form**, the standard forms recommended in CDM are first shown, and then their pole locations and step responses are presented. In **Robustness Consideration**, the mechanism, whereby the robustness is lost, is first explained, and some guidelines are given to modify the standard forms to recover robustness. By proper use of the basic knowledge introduced in this chapter, the designers will be able to proceed in the CDM design efficiently. In **Summary**, the important points in this chapter are summarized.

2.1 Mathematical Relations

In this section, important terms and symbols, commonly used in CDM, are first defined. Then mathematical relations among these symbols are derived. Such mathematical relations are very useful in CDM design. The **characteristic polynomial** $P(s)$ is given in the following form:

$$P(s) = a_n s^n + a_{n-1} s^{n-1} + \cdots + a_1 s + a_0 = \sum_{i=0}^{n} a_i s^i. \tag{2.1}$$

The **stability index** γ_i, the **equivalent time constant** τ, and the **stability limit** γ_i^* are defined as follows:

S. Manabe and Y. C. Kim, *Coefficient Diagram Method for Control System Design*, Intelligent Systems, Control and Automation: Science and Engineering 99,
https://doi.org/10.1007/978-981-16-0546-8_2

$$\gamma_i = a_i^2/(a_{i+1}a_{i-1}), \quad i = 1 \cdots n-1. \tag{2.2}$$

$$\tau = a_1/a_0. \tag{2.3}$$

$$\gamma_i^* = 1/\gamma_{i+1} + 1/\gamma_{i-1}, \; i = 1 \cdots n-1, \; \gamma_n = \gamma_0 = \infty. \tag{2.4}$$

It has been explained already that the stability index is the measure of stability, and the equivalent time constant is the measure of response speed. The stability limit is the newly introduced term and is used to express stability more accurately. In the equation, γ_n and γ_0 are defined as ∞. By this definition, Eq. (2.4) can be applied to γ_{n-1}^* and γ_1^* without difficulty. Because the coefficients a_{n+1} and a_{-1} in the nth order characteristic polynomial are naturally considered as 0, these definitions are consistent with Eq. (2.2). By generalizing the concept of the equivalent time constant, the i**th order equivalent time constant** τ_i is defined as follows:

$$\tau_i = a_{i+1}/a_i, \quad i = 1 \cdots n-1. \tag{2.5}$$

From Eqs. (2.2) and (2.3), the following relations are obtained.

$$\tau_i = \tau_{i-1}/\gamma_i = \tau/(\gamma_i \cdots \gamma_2\gamma_1). \tag{2.6}$$

$$a_i = \tau_{i-1} \cdots \tau_2\tau_1\tau a_0 = a_0\tau^i/(\gamma_{i-1}\gamma_{i-2}^2 \cdots \gamma_2^{i-2}\gamma_1^{i-1}). \tag{2.7}$$

As the result, the **characteristic polynomial** is expressed by a_0, τ, and γ_i in the following form:

$$P(s) = a_0 \left[\left\{ \sum_{i=2}^{n} \left(\prod_{j=1}^{i-1} 1/\gamma_{i-j}^{j} \right) (\tau s)^i \right\} + \tau s + 1 \right]. \tag{2.8}$$

It is to be noticed that the characteristic polynomial consists of the powers of (τs) and coefficients which are expressed by stability indices. This shows that the response waveform is completely defined by stability indices and the time scale of the response is given by the equivalent time constant. From Eqs. (2.5) and (2.6), the following useful **formula for adjacent coefficients** are derived:

$$\frac{a_i a_j}{a_{i+1}a_{j-1}} = \gamma_i\gamma_{i-1} \cdots \gamma_j, \quad i > j. \tag{2.9}$$

Also, the **stability index of the second order** is defined as follows:

$$\gamma_{i2} = a_i^2/(a_{i+2}a_{i-2}) = \gamma_{i+1}\gamma_i^2\gamma_{i-1}, \quad i = 2 \cdots n-2. \tag{2.10}$$

This index is effective in making rough estimation of stability. From Eq. (2.2), a relation concerning the **ratio of stability indices** is obtained as follows:

$$\gamma_{i-1}/\gamma_{i+1} = (a_{i-1}/a_{i+1})^2/(a_{i-2}/a_{i+2}), \quad i = 2 \cdots n-2. \tag{2.11}$$

This relation shows the degree of symmetry of the curvature of the diagram about a_i. When the value is 1, the curvature of the diagram is in complete symmetry. In this situation, the right hand side of the equation shows that the line connecting a_{i+1} and a_{i-1} is in parallel with the line connecting a_{i+2} and a_{i-2}. The coefficient diagram becomes smooth and balanced. Such characteristic polynomial usually shows favorable control characteristics in many respects. Several **coefficient relations** useful in design will be shown hereafter. Corresponding to Eq. (2.7), the equation to obtain a_0 from a_i is shown as follows:

$$a_0 = a_i/(\tau_{i-1}\cdots\tau_2\tau_1\tau) = \frac{a_i}{(\tau_{i-1}^i)(\gamma_{i-1}^{i-1}\gamma_{i-2}^{i-2}\cdots\gamma_2^2\gamma_1)}. \tag{2.12}$$

This equation is very convenient, when the total characteristic polynomial is to be constructed out of the information of the high-order terms. Equations (2.7) and (2.12) are generalized in the following forms:

$$a_{i+j}/a_i = \tau_{i+j-1}\tau_{i+j-2}\cdots\tau_i = \frac{\tau_i^j}{\gamma_{i+j-1}\gamma_{i+j-2}^2\cdots\gamma_{i+1}^{j-1}}. \tag{2.13}$$

$$a_{i-j}/a_i = 1/(\tau_{i-1}\tau_{i-2}\cdots\tau_{i-j}) = \frac{1}{(\tau_{i-1}^j)(\gamma_{i-1}^{j-1}\gamma_{i-2}^{j-2}\cdots\gamma_{i-j+1})}. \tag{2.14}$$

By use of these equations, the **jth order stability index** is expressed as follows:

$$\begin{aligned}\gamma_{ij} &= a_i^2/(a_{i+j}a_{i-j}) \\ &= (\gamma_{i+j-1}\gamma_{i-j+1})\cdots(\gamma_{i+1}\gamma_{i-1})^{j-1}\gamma_i^j, \quad i+j \le n, \quad i-j \ge 0.\end{aligned} \tag{2.15}$$

Finally, the relation between the stability index γ_i and the corresponding **damping ratio** ζ_i is shown. The detail will be discussed in the later chapter, but the following approximation usually holds:

$$\zeta_i \simeq 0.5\sqrt{\gamma_i}(1 - \gamma_i^*/\gamma_i). \tag{2.16}$$

Thus, ζ_i becomes $0.5\sqrt{\gamma_i}$, When γ_i^*/γ_i is small and the corresponding portion of the characteristic polynomial is close to the second order. Otherwise, ζ_i is smaller than $0.5\sqrt{\gamma_i}$.

2.2 Coefficient Diagram

In this section, the coefficient diagram, which plays a crucial role in CDM, will be explained. **Structure of Coefficient Diagram** is first discussed. It will be shown that the coefficient diagram of the characteristic polynomial is composed of several component polynomials. Then a simple example of **Coefficient Diagram** and its drawing procedures are shown. Also, the way to read the three key elements of control systems, namely stability, response, and robustness, from the diagram is explained. Finally, the **Unrobust Structure** will be discussed. Generally speaking, the design by CDM is inherently robust, and the need to consider robustness is low. But in some special problems, robustness has to be considered. In such case, the coefficient diagram shows some strange features and the problem is easily identified. Also, the relation with the robust control theory will be briefly discussed.

The **Structure of Coefficient Diagram** will be explained hereafter. As explained in Sect. 1.4, the characteristic polynomial $P(s)$ is the sum of the denominator polynomial $A_{cp}(s)$ and the numerator polynomial $B_{cp}(s)$ of the open-loop transfer function $G(s)$.

$$P(s) = A_{cp}(s) + B_{cp}(s), \quad G(s) = B_{cp}(s)/A_{cp}(s), \tag{2.17}$$

where

$$A_{cp}(s) = A_c(s)A_p(s), \quad B_{cp}(s) = B_c(s)B_p(s).$$

In order to simplify the explanation, the order of the controller is set as 1/1 (where, order 1 for numerator and order 1 for the denominator).

$$A_{cp}(s) = (l_1 s + l_0)A_p(s), \quad B_{cp}(s) = (k_1 s + k_0)B_p(s). \tag{2.18}$$

Then

$$P(s) = P_{l_1}(s) + P_{l_0}(s) + P_{k_1}(s) + P_{k_0}(s), \tag{2.19}$$

where

$$\begin{aligned} P_{l_1}(s) &= l_1 s A_p(s), \quad P_{l_0}(s) = l_0 A_p(s), \\ P_{k_1}(s) &= k_1 s B_p(s), \quad P_{k_0}(s) = k_0 B_p(s). \end{aligned}$$

On the coefficient diagram, $P_{l_1}(s)$ is obtained by moving $A_p(s)$ by 1 order to the left, and moving up and down by the amount of l_1. $P_{l_0}(s)$ is obtained by moving $A_p(s)$ up and down by the amount of l_0. $P_{k_1}(s)$ and $P_{k_0}(s)$ are obtained by the same operation to $B_p(s)$. Thus, the characteristic polynomial $P(s)$ is composed of 4 component polynomials generated by the horizontal and vertical movement of $A_p(s)$ and $B_p(s)$. For the general order controller, not 1/1 order, the characteristic polynomial is composed of component polynomials, whose number is equal to that of controller parameters. When the characteristic polynomial is decomposed to such component polynomials, the specific component polynomials, which give large influence to the characteristic polynomial, are easily identified. Then the controller parameters, which are most effective in parameter tuning, can be easily identified. When the number of component polynomials is too large, it is difficult to obtain the total image. Then it is convenient to summarize these polynomials to 2 polynomials, namely the denominator component polynomial $P_l(s)$ and the numerator component polynomial $P_k(s)$.

$$P_l(s) = P_{l_1}(s) + P_{l_0}(s) = A_{cp}(s), \tag{2.20}$$

$$P_k(s) = P_{k_1}(s) + P_{k_0}(s) = B_{cp}(s). \tag{2.21}$$

This expression is convenient in the verification of robustness. Also, it is used in the prediction of the frequency response and step response as will be shown in the later chapter.

By a simple example of **Coefficient Diagram**, its drawing procedures will be first shown, and then the way to read stability, response, and robustness from the diagram will be explained. The denominator/numerator polynomials of the plant are given as follows:

$$A_p(s) = 0.25s^4 + s^3 + 2s^2 + 0.5s, \quad B_p(s) = 1. \tag{2.22}$$

A 2/1 order controller (PID controller), as shown below, is attached to this plant.

$$A_c(s) = l_1 s, \quad B_c(s) = k_2 s^2 + k_1 s + k_0, \tag{2.23}$$

where

$$l_1 = 1, \quad k_2 = 1.5, \quad k_1 = 1, \quad k_0 = 0.2.$$

The component polynomials are as follows:

$$P_l(s) = A_{cp}(s) = 0.25s^5 + s^4 + 2s^3 + 0.5s^2, \tag{2.24}$$

$$P_k(s) = B_{cp}(s) = 1.5s^2 + s + 0.2. \tag{2.25}$$

The characteristic polynomial becomes as follows:

$$P(s) = 0.25s^5 + s^4 + 2s^3 + 2s^2 + s + 0.2. \tag{2.26}$$

Then according to the definitions, the following results are obtained:

$$a_i = [a_5 \; \cdots \; a_2 \; a_1] = [0.25 \;\; 1 \;\; 2 \;\; 2 \;\; 1 \;\; 0.2], \tag{2.27}$$

$$\gamma_i = [\gamma_4 \; \cdots \; \gamma_2 \; \gamma_1] = [2 \;\; 2 \;\; 2 \;\; 2.5], \tag{2.28}$$

$$\tau = 5, \tag{2.29}$$

$$\gamma_i^* = [\gamma_4^* \; \cdots \; \gamma_2^* \; \gamma_1^*] = [0.5 \;\; 1 \;\; 0.9 \;\; 0.5]. \tag{2.30}$$

In the above expression, a_i is a row vector, and the suffix is in descending order. This descending order convention corresponds to the descending power order of the characteristic polynomial. This convention is adopted throughout in CDM, and γ_i and γ_i^* are expressed in this way.

The coefficient diagram is shown in Fig. 2.1. The coefficient a_i is read by the left hand scale, and the stability index γ_i, the equivalent time constant τ, and the stability limit γ_i^* are read by the right hand scale. The τ is represented by a line, which connects the point 1 at $i = 0$ and the point τ at $i = 1$. This line is obtained

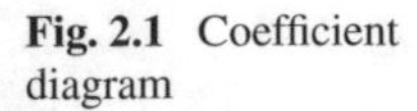

Fig. 2.1 Coefficient diagram

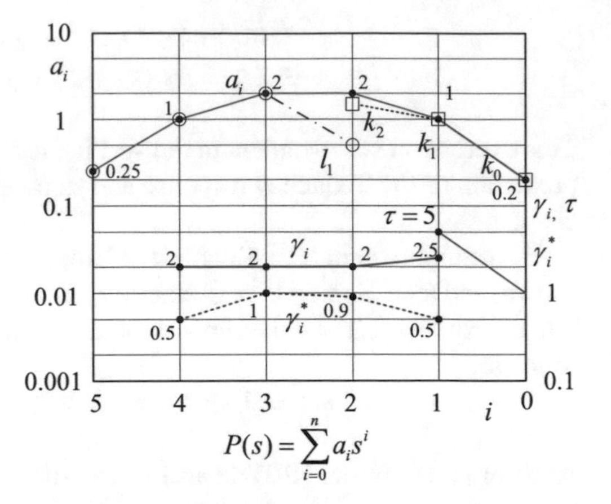

Fig. 2.2 Effect of γ_i

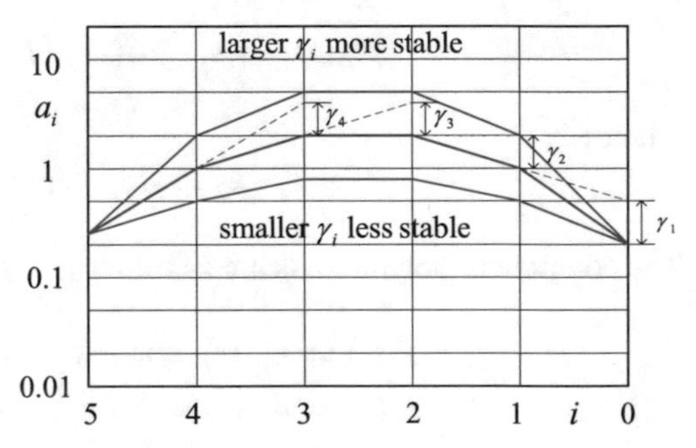

by the parallel movement of the line connecting a_1 and a_0. The coefficients, stability indices, and equivalent time constant are represented by the dots connected by solid lines. The stability limits are represented by the dots connected by dashed lines. The $P_l(s)$ is represented by the circles and dash-dot lines, and the $P_k(s)$ by squares and dotted lines.

The stability index can be obtained graphically as in Fig. 2.2. In order to obtain γ_i, just extend the line connecting a_{i+1} and a_i to $i-1$, and read the difference of the extended point and a_{i-1} in the logarithmic scale. This shows that the stability index is proportional to the curvature of the coefficient diagram of the characteristic polynomial. As shown in Fig. 2.2, when the curvature becomes large, the stability index γ_i becomes large, and stability is increased. In this way, the stability of the system can be estimated by the curvature of the coefficient diagram of the characteristic polynomial.

As in Fig. 2.3, when the coefficient diagram of the characteristic polynomial is left-end-down, the equivalent time constant τ becomes small, and the response becomes fast. In this way, the response speed of the system can be estimated by the inclination of the coefficient diagram of the characteristic polynomial.

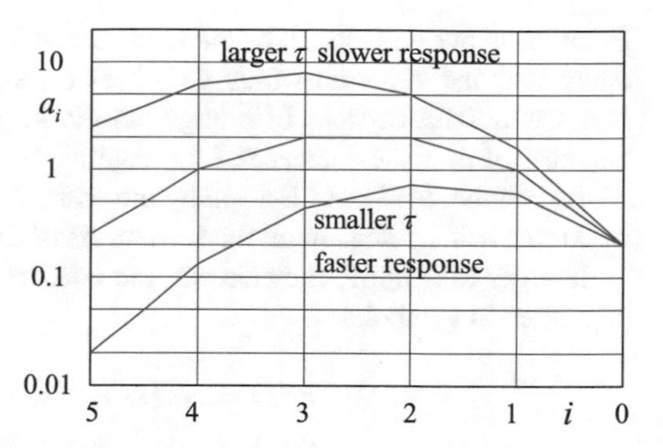

Fig. 2.3 Effect of τ

Many definitions of "robustness" are currently in use. In CDM, the robustness is defined as "the robustness of the stability for the variation of the specific plant/controller parameters". When the major portion of the coefficient diagram of the characteristic polynomial does not vary too much for the variation of such parameters, the robustness is very high. When the appreciable variation is experienced in the coefficient diagram, the robustness is very low. The parameter variation first appears at the variation of the specific component polynomial, such as $P_l(s)$ or $P_k(s)$ in this example. Then the variation of the coefficient diagram of the characteristic polynomial becomes visible. In CDM, the robustness can be easily estimated, because the variation of the characteristic polynomial due to such parameter variation can be read visually at a glance in the coefficient diagram. As explained above, the robustness is addressed to the variation of the specific parameter in CDM. In the terminology of the robust control theory, "structured uncertainty" is in question in CDM.

The **Unrobust Structure** will be discussed hereafter. In CDM design, sufficient robustness is inherently guaranteed in the usual case. However, in some special structures, the robustness becomes low. Such special structures will be identified in the following analysis.

In the standard characteristic polynomial recommended in CDM, the robustness to the variation of the single coefficient is very high. Generally speaking, as will be made clear in the later chapter, the system is stable for the variation of approximately from 0.5 to 3 times of the coefficient. Furthermore, the lower bound is extended to 0 times for a_n and a_0, and approximately 0.25 times for a_{n-1} and a_1. The allowable variation δa_i of the coefficient a_i is approximately given by the following equation:

$$0a_i(i=n,0),\ 0.25a_i(i=n-1,1),\ 0.5a_i(n-2\geq i\geq 2) < a_i+\delta a_i < 3a_i. \tag{2.31}$$

In terms of the ratio of variation $\delta a_i/a_i$, the stability limit of the variation is expressed approximately by the following equation:

$$-1(i=n,0),\ -0.75(i=n-1,1),\ -0.5(n-2\geq i\geq 2) < \frac{\delta a_i}{a_i} < 2. \tag{2.32}$$

From the above equation, it becomes clear that the important variation for robustness analysis is the variation which decreases a_i, and the variation which increases a_i is not important because of a large margin in that direction. One cause of such a decrease of a_i is the decrease of the negative feedback, and the other is the increase of the positive feedback. Especially the existence of positive feedback plays a very important role in weakening the robustness of the system.

In order to simplify the analysis, the coefficients of component polynomials are expressed in symbols.

$$P_l(s) = c_5 s^5 + c_4 s^4 + c_3 s^3 + c_2 s^2, \tag{2.33}$$

$$P_k(s) = b_2 s^2 + b_1 s + b_0. \tag{2.34}$$

In this example, all the coefficients of the characteristic polynomial, except a_2, are equal to those of component polynomial. The ratio of variation of each coefficient of the characteristic polynomial, except a_2, is equal to that of the component polynomial.

For a_2, the following relation is obtained:

$$\begin{aligned}\frac{\delta a_2}{a_2} &= (c_2/a_2)\frac{\delta c_2}{c_2} + (b_2/a_2)\frac{\delta b_2}{b_2}, \\ &= (0.5/2)\frac{\delta c_2}{c_2} + (1.5/2)\frac{\delta b_2}{b_2}, \\ &= 0.75\frac{\delta b_2}{b_2}.\end{aligned} \tag{2.35}$$

In this equation, the parameter variation was assumed only for b_2. The ratio of variation of the coefficient of the characteristic polynomial is smaller than that of the component polynomial. If $\delta a_2/a_2$, corresponding to the stability limit, is considered to be -0.5, then the corresponding $\delta b_2/b_2$ becomes -0.6667 with larger stability margin. As shown in the coefficient diagram of Fig. 2.1, the point b_2 is below the point a_2. In such circumstances, the effect of the parameter variation of the component polynomial is reduced, and the robustness of the system is increased.

If the points representing the coefficients of the component polynomials are below those of the characteristic polynomial, the ratio of variation of the coefficients of the characteristic polynomial is smaller than those of component polynomials. Then the sufficient robustness is retained. In this example, the ratios of variation of the coefficients of the characteristic polynomial are equal or less than those of component polynomials, and robustness is sufficient.

The next example shows the case of low robustness. The component polynomials are changed as follows; the characteristic polynomial does not change.

$$P_l(s) = 0.25s^5 + s^4 + 2s^3 + 6s^2, \tag{2.36}$$

$$P_k(s) = -4s^2 + s + 0.2. \tag{2.37}$$

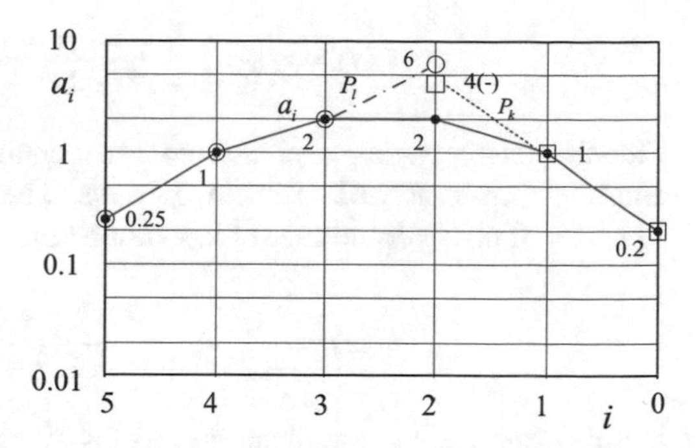

Fig. 2.4 Unrobust example

The ratio of variation of a_2 corresponding to b_2 increases appreciably as follows;

$$\begin{aligned}\frac{\delta a_2}{a_2} &= (c_2/a_2)\frac{\delta c_2}{c_2} + (b_2/a_2)\frac{\delta b_2}{b_2}, \\ &= (6/2)\frac{\delta c_2}{c_2} + (-4/2)\frac{\delta b_2}{b_2}, \\ &= -2\frac{\delta b_2}{b_2}. \end{aligned} \tag{2.38}$$

The ratio of variation of the coefficient of the characteristic polynomial is twice as large as that of component polynomial, and the robustness becomes low. If $\delta a_2/a_2$, corresponding to the stability limit, is considered to be -0.5, then the corresponding $\delta b_2/b_2$ becomes 0.25 with much smaller stability margin. Because $b_2 = -4$, the point is located above the point of $a_2 = 2$ in the coefficient diagram. In CDM convention, the negative coefficient is shown in the absolute value with (−) indication in the coefficient diagram. Thus the cause of poor robustness is that the coefficient, in absolute value, of the component polynomial is located above the corresponding coefficient of the characteristic polynomial.

In such a situation, the coefficient of the characteristic polynomial is composed of coefficients of component polynomials with different signs. The coefficient of the characteristic polynomial becomes the difference of two large numbers. The existence of such structure, characterized by the difference of two large numbers, is the fundamental structure of the unrobust system. It is indicated by the coefficient of the component polynomial located above that of the characteristic polynomial as in Fig. 2.4.

Finally, the relation between "structured uncertainty" used in CDM and "unstructured uncertainty" used in robust control theories will be briefly explained with an example. The complementary sensitivity function $T(s)$ is shown as follows:

$$T(s) = \frac{P_k(s)}{P(s)}. \tag{2.39}$$

In the previous example, Eqs. (2.25) and (2.26) result in

$$T(s) = \frac{1.5s^2 + s + 0.2}{0.25s^5 + s^4 + 2s^3 + 2s^2 + s + 0.2}. \tag{2.40}$$

The frequency response of this complementary sensitivity function shows the amplitude peak of 1.61 at about 0.8 rad/s. The permissible uncertainty becomes $1/1.61 = 0.621$. For the case of low robustness such as in Eq. (2.37),

$$T(s) = \frac{-4s^2 + s + 0.2}{0.25s^5 + s^4 + 2s^3 + 2s^2 + s + 0.2} \tag{2.41}$$

In this example, the amplitude peak of 4.19 is indicated at about 0.8 rad/s. The permissible uncertainty becomes small as $1/4.19 = 0.238$. It is noticeable that the permissible uncertainties by the robust control theories are similar to the permissible variations -0.6667 and 0.25 of $\delta b_2/b_2$ by CDM analysis. By CDM, a more detailed analysis of robustness is possible. However, it also includes the result of robust control theories. In robust control theories, it is understood that the amplitude of the complementary sensitivity function is large when robustness is low. This corresponds to the case that the coefficient of the component polynomial is located above the coefficient of the characteristic polynomial in the coefficient diagram.

As explained above, the coefficient diagram of the characteristic polynomial shows the stability of the system by its curvature, the speed of response by its inclination, and the robustness by the comparison with the component polynomial. The most important feature of the coefficient diagram is that the three key elements of the control system are shown in a single diagram. The coefficient diagram may replace the Bode diagram in future, because it is more effective and accurate at the design stage.

2.3 Stability Condition

In this section, the sufficient conditions for stability and instability by Lipatov are explained. They are the simplification of the Routh stability condition and are very useful in CDM design. Especially the sufficient condition for stability is most useful, and constitutes the theoretical basis of CDM. First, **Routh Stability Condition**, whose expression is somewhat modified to fit CDM, is explained. Then **Third and Fourth-Order Routh Stability Conditions** are shown and their graphical expressions on coefficient diagrams are explained. Then **Sufficient Conditions for Stability and Instability by Lipatov** are explained. They apply to the systems equal or higher than fifth order. Although the Lipatov's sufficient conditions may be conservative in some cases, but their simplicity makes it very convenient for stability analysis and controller design. Finally, some **Examples** are shown to help to understand the stability conditions.

Routh Stability Condition In this section, the Routh stability condition is explained in the form slightly modified from the standard expression. By such modification, the expression becomes compatible with the other CDM expression. The nth-order characteristic polynomial $P_n(s)$ is first given. The order n is assumed as an odd number for convenience. The coefficients of the characteristic polynomial are all positive, because otherwise it is immediately found to be unstable.

$$P_n(s) = a_n s^n + a_{n-1} s^{n-1} + a_{n-2} s^{n-2} + \cdots + a_3 s^3 + a_2 s^2 + a_1 s + a_0. \quad (2.42)$$

From this equation, the $(n-1)$th-order characteristic polynomial is obtained in the following manner:

$$\begin{aligned} P_{n-1}(s) = {} & a_{n-1} s^{n-1} + a'_{n-2} s^{n-2} + a_{n-3} s^{n-3} + \cdots + a_4 s^4 + a'_3 s^3 \\ & + a_2 s^2 + a'_1 s + a_0, \end{aligned} \quad (2.43)$$

where

$$a'_i = a_i - a_n(a_{i-1}/a_{n-1}), \quad i = n-2, \quad \cdots \quad 3, \quad 1.$$

The proof of Routh stability condition basically states that $P_n(s)$ and $P_{n-1}(s)$ are equivalent in stability. If one is stable, the other is stable, and vice versa. If this process is repeated, the second order polynomial $P_2(s)$ is finally obtained. All coefficients of all polynomials, which appear in this process is assumed to be positive. If zero or negative coefficients appear in some polynomial, that polynomial is unstable, and eventually the original characteristic polynomial $P_n(s)$ is found to be unstable. Because $P_2(s)$ is second order, its stability is assured, if the coefficients are all positive. Thus, if coefficients of polynomials $P_{n-1}(s)$ through $P_2(s)$ are all positive, The original characteristic polynomial $P_n(s)$ is assured to be stable. In the above development, the order n is assumed as an odd number. If it is even, it suffices to modify Eq. (2.43) in the following manner:

$$\begin{aligned} P_{n-1}(s) = {} & a_{n-1} s^{n-1} + a'_{n-2} s^{n-2} + a_{n-3} s^{n-3} + \cdots + a'_4 s^4 + a_3 s^3 \\ & + a'_2 s^2 + a_1 s + a_0, \end{aligned} \quad (2.44)$$

where

$$a'_i = a_i - a_n(a_{i-1}/a_{n-1}), \quad i = n-2 \quad \cdots \quad 4 \ 2.$$

The process of building $P_{n-1}(s)$ from $P_n(s)$ is illustrated in the coefficient diagram of the fifth-order characteristic polynomial as in Fig. 2.5. Draw a straight line from a_5 in parallel with the line $a_4 a_2$. The point, where the line crosses with $i = 3$, is $a_5(a_2/a_4)$. Subtract this value from a_3. The result is a'_3. Similarly draw a straight line from a_5 in parallel with the line $a_4 a_0$. The point, where the line crosses with $i = 1$, is $a_5(a_0/a_4)$. Subtract this value from a_1. The result is a'_1. By the first glance of the coefficient diagram, it is clear that a'_3 is somewhat smaller than a_3, but a'_1 is almost

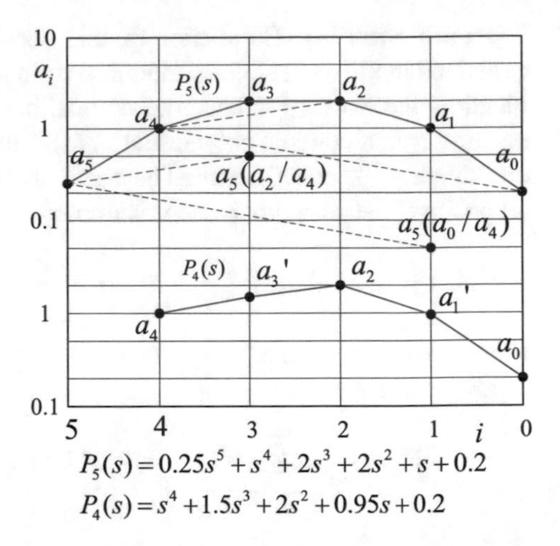

Fig. 2.5 Routh stability condition

equal to a_1. When the Routh stability condition is expressed in such a graphical manner, the condition can be easily remembered.

Routh Stability Condition for Third and Fourth-Order Systems will be discussed next. For the third-order system, $P_2(s)$ is obtained from $P_3(s)$ as follows:

$$P_3(s) = a_3 s^3 + a_2 s^2 + a_1 s + a_0, \tag{2.45}$$

$$P_2(s) = a_2 s^2 + (a_1 - a_3(a_0/a_2))s + a_0. \tag{2.46}$$

Thus, the stability condition is $a_1 - a_3(a_0/a_2) > 0$, that is

$$a_2 a_1 > a_3 a_0. \tag{2.47}$$

Since $a_2 a_1/(a_3 a_0) = \gamma_2 \gamma_1$, the equation is expressed in terms of stability indices as follows:

$$\gamma_2 \gamma_1 > 1. \tag{2.48}$$

For the fourth-order system, $P_3(s)$ is obtained from $P_4(s)$ as follows:

$$P_4(s) = a_4 s^4 + a_3 s^3 + a_2 s^2 + a_1 s + a_0, \tag{2.49}$$

$$P_3(s) = a_3 s^3 + (a_2 - a_4(a_1/a_3))s^2 + a_1 s + a_0. \tag{2.50}$$

By the third-order stability condition, the ineqality, $(a_2 - a_4(a_1/a_3))a_1 > a_0 a_3$, is obtained. After rearrangement, the following fourth-order stability condition is obtained.

$$a_2 > a_4 \left(\frac{a_1}{a_3}\right) + a_0 \left(\frac{a_3}{a_1}\right) \tag{2.51}$$

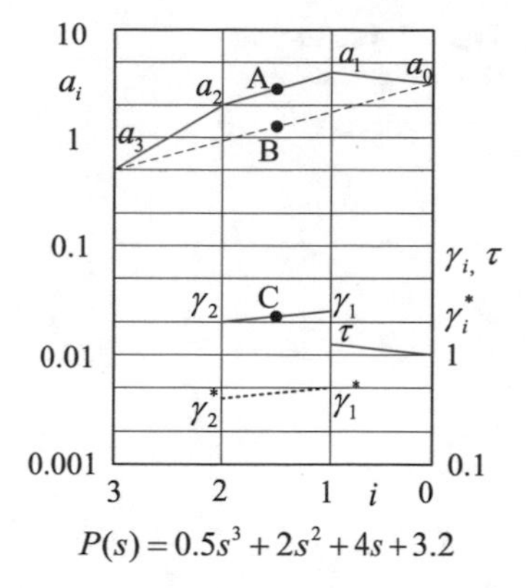

Fig. 2.6 Third-order stability condition

This condition can be expressed in terms of stability indices. Multiply both sides by a_2 and divide by (a_3a_1). Then the equation becomes $a_2^2/(a_3a_1) > (a_4a_2)/a_3^2 + (a_2a_0)/a_1^2$. By the definitions of stability index and stability limit, the following relation is obtained:

$$\gamma_2 > 1/\gamma_3 + 1/\gamma_1 = \gamma_2^* \tag{2.52}$$

The third and fourth-order stability conditions can be expressed graphically on the coefficient diagram. Figure 2.6 is an example for the third-order system. The point A represents $(a_2\ a_1)^{0.5}$, and the point B is $(a_3\ a_0)^{0.5}$. If the point A is above the point B, the system is stable. The point C is $(\gamma_2\ \gamma_1)^{0.5}$. Thus, the system is stable because it is above 1.

Figure 2.7 is an example of the fourth-order system. Draw a straight line from a_4 in parallel with the line a_3a_1. Find the crossing point A of the line with $i = 2$. Similarly, draw a straight line from a_0 in parallel with the line a_3a_1. Find the crossing point B of the line with $i = 2$. The stability condition is $a_2 > (A + B)$. The stability condition in terms of stability index and stability limit, $\gamma_2 > \gamma_2^*$, can be read directly from the diagram. In this way, the Routh stability conditions for third and fourth order are graphically expressed on the coefficient diagram in a much understandable form.

Sufficient Conditions for Stability and Instability by Lipatov apply to the systems equal or higher than fifth order. These stability criteria were proposed by Lipatov and Sokolov in 1978 [1]. Because they are only sufficient conditions and not exact stability conditions, some ambiguity remains in determining the stability. However, because of their simplicity, they are very useful in the actual design. (For more details, refer to the literature [2]). There are several expressions for stability conditions by Lipatov. However, the expression most suitable to CDM is as follows:

> The system is stable if all partial fourth-order polynomials of the characteristic polynomial are stable with the margin of 1.12. The system is unstable if some partial third-order polynomial is unstable.

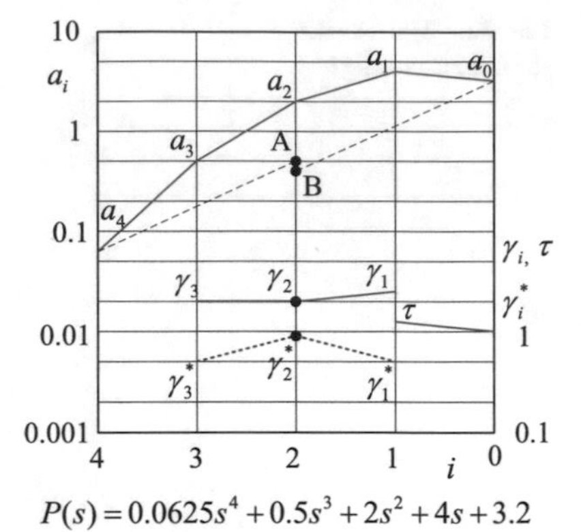

Fig. 2.7 Fourth-order stability condition

Thus, the sufficient condition for stability is as follows:

$$a_i > 1.12 \left[a_{i+2} \left(\frac{a_{i-1}}{a_{i+1}} \right) + a_{i-2} \left(\frac{a_{i+1}}{a_{i-1}} \right) \right], \tag{2.53}$$

$$\gamma_i > 1.12\gamma_i^*, \quad \text{for all } i = n-2 \cdots 2. \tag{2.54}$$

The sufficient codition for instability is as follows:

$$a_{i+1}a_i < a_{i+2}a_{i-1}, \tag{2.55}$$

$$\gamma_{i+1}\gamma_i < 1, \quad \text{for some } i = n-2 \cdots 1. \tag{2.56}$$

Lipatov first proved that the system is stable if all partial fifth-order polynomials of the characteristic polynomial are stable. Then expression is simplified by replacement of the fifth-order stability condition by the sufficient condition for stability of fourth order. In this process, a different expression is proposed. The present expression is chosen because it is most suitable to CDM. The number 1.12 is approximate value and the exact value is $(3/4^{1/3} - 1)^{-1} = 1.1237$. Lipatov proposed a more accurate expression for the sufficient condition for instability. The present expression is adopted from the standpoint of practicality. In ordinary characteristic polynomials, experiences show that the stability of all partial fourth-order polynomials can be used as the stability condition of the total system.

Finally, **Examples** are shown to help understanding of the stability conditions by Lipatov. Figure 2.8 is a sixth-order example from Franklin's text book [3, see p. 171], [4, p. 217]. The characteristic polynomial is as follows:

$$P(s) = s^6 + 4s^5 + 3s^4 + 2s^3 + s^2 + 4s + 4. \tag{2.57}$$

At first glance, the worst points can be found at the four adjacent coefficient segments of $[a_4a_3a_2a_1]$. Because $A < B$, the sufficient condition for instability is satisfied, and the system is unstable. Figure 2.9 is a fifth-order example from the same textbook

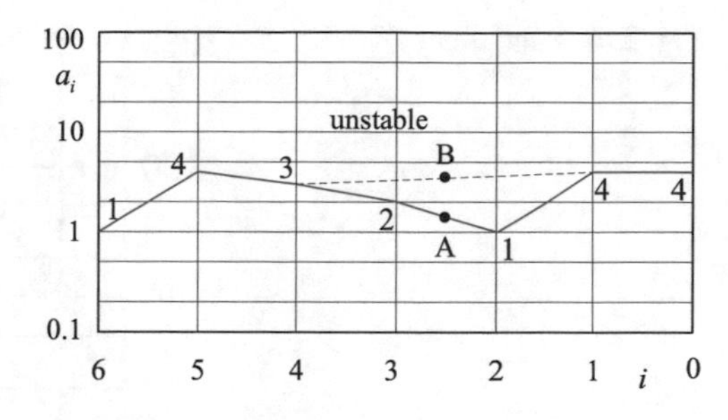

Fig. 2.8 Sixth-order example

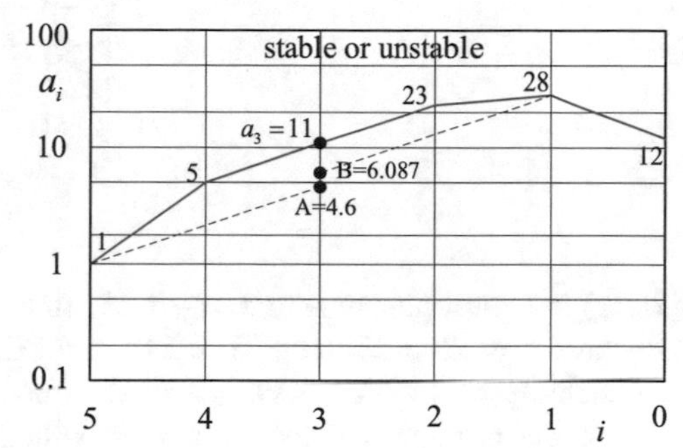

Fig. 2.9 Fifth-order example

[4, p. 219]. The characteristic polynomial is as follows:

$$P(s) = s^5 + 5s^4 + 11s^3 + 23s^2 + 28s + 12. \tag{2.58}$$

At first glance, the worst point is $a_3 = 11$. Because $A = 23/5 = 4.6$, $B = 28 \times (5/23) = 6.087$, $A + B = 10.687$, and $11/10.687 = 1.0293$, the sufficient condition for stability is not satisfied. Also, by looking at the diagram, it is clear that the sufficient condition for instability is not satisfied either. In fact, this system is on the boundary of stability with imaginary roots at $\pm j2$. It is very interesting to note that $(a_2/a_4)^{0.5} = 2.145$ is approximately equal to these imaginary roots. From this example, it is observed that the stability of all partial fourth-order polynomials is virtually equivalent to the stability of the total system.

Let us consider two tenth order examples as shown in Fig. 2.10. Each characteristic polynomial $P_A(s)$ and $P_B(s)$ is given as follows:

$$\begin{aligned} P_A(s) &= (s+1)^{10} \\ &= s^{10} + 10s^9 + 45s^8 + 120s^7 + 210s^6 + 252s^5 \\ &\quad + 210s^4 + 120s^3 + 45s^2 + 10s + 1. \end{aligned} \tag{2.59}$$

$$\begin{aligned} P_B(s) &= (s+1)^{10} - 32s^5 \\ &= s^{10} + 10s^9 + 45s^8 + 120s^7 + 210s^6 + 220s^5 \\ &\quad + 210s^4 + 120s^3 + 45s^2 + 10s + 1. \end{aligned} \tag{2.60}$$

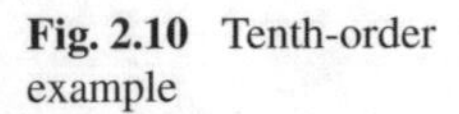
Fig. 2.10 Tenth-order example

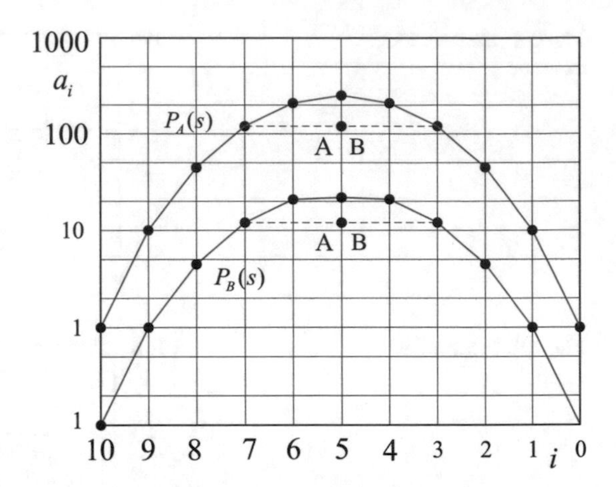

$P_A(s)$ has multiple roots at $s = -1$. At first glance, the worst point is $a_5 = 252$. Because $A = B = 120$, $A + B = 240$, and $252/240 = 1.05$, the sufficient condition for stability by Lipatov is not satisfied, although the system is very stable. Proper precaution is requested for the system with multiple roots because the sufficient condition for stability by Lipatov is not effective. However, such a case is rare.

$P_B(s)$ is obtained by modifying $P_A(s)$ in such a manner that it stays on the stability boundary. By simple calculation, $\pm j$ are found to be the roots. The worst point is $a_5 = 220$. Because $A = B = 120$, $A + B = 240$, and $220/240 = 0.91667$, not only the sufficient condition for stability by Lipatov, but also the fourth-order stability condition is not satisfied. When the coefficient is modified by an infinitesimally small amount to the stability side, the system is stable, but the partial fourth-order polynomial does not satisfy the stability condition. In other words, the instability of the partial fourth-order polynomial cannot be the sufficient condition for instability of the total system. In the above discussion, the application of the sufficient conditions for stability and instability by Lipatov is shown with some examples. A more detailed discussion will be made in later chapters with specific design examples.

2.4 Canonical Transfer Functions

In this section, the canonical transfer function is newly defined and introduced. By this function, the relation between the characteristic polynomial and the general transfer function will be clarified. The transfer function is composed of the denominator and the numerator. The denominator is given as the characteristic polynomial, but the numerator must be defined by other means to define the transfer function. For this reason, some special means are necessary to relate the characteristic polynomial to the transfer function. First, **Canonical Transfer Function** is defined. This function is the special transfer function solely defined by the characteristic polynomial.

All transfer functions can be expressed by the combination of canonical transfer functions. Second, **System Type** is explained, and the transfer functions related to such system types are obtained. In this process, the relation between the classical control, where the open-loop transfer function is mainly used, and CDM, where the characteristic polynomial is the main concern, is clarified. Finally, **CDM-type Bode Diagram** is defined, and its relation with the coefficient diagram will be clarified.

Canonical Transfer Function is defined in the following manner. The characteristic polynomial $P(s)$ is given as follows:

$$P(s) = a_n s^n + a_{n-1} s^{n-1} + \cdots + a_1 s + a_0. \tag{2.61}$$

Then the ith order canonical transfer function $T_i(s)$ is defined as follows:

$$T_i(s) = \frac{a_i s^i}{a_n s^n + a_{n-1} s^{n-1} + \cdots + a_1 s + a_0}. \tag{2.62}$$

With these equations, the general closed-loop transfer function $F(s)$ is expressed as follows:

$$\begin{aligned} F(s) &= \frac{b_m s^m + b_{m-1} s^{m-1} + \cdots + b_1 s + b_0}{a_n s^n + a_{n-1} s^{n-1} + \cdots + a_1 s + a_0}, \\ &= \sum_{i=o}^{m} (b_i/a_i) T_i(s). \end{aligned} \tag{2.63}$$

Also the sum of two canonical transfer functions, $T_i(s)$ and $T_j(s)$, is called the $[i, j]$th order canonical transfer function $T_{i,j}(s)$.

$$T_{i,j}(s) = T_i(s) + T_j(s). \tag{2.64}$$

So far, the closed-loop transfer functions are treated. Now the open-loop transfer function is considered. When a closed-loop transfer function is given, there exist infinite number of open-loop transfer functions. But if the unity feedback structure is assumed, only one open-loop transfer function is defined. Such open-loop transfer function corresponding to the canonical transfer function is called canonical open-loop transfer function, and expressed as $G_i(s)$ or $G_{i,j}(s)$. They are related to the canonical transfer functions as follows:

$$G_i(s) = T_i(s)/[1 - T_i(s)]. \tag{2.65}$$

$$G_{i,j}(s) = T_{i,j}(s)/[1 - T_{i,j}(s)]. \tag{2.66}$$

Thus, so far the definitions of canonical transfer functions are given and related matters are discussed.

System Type is a concept in classical control introduced in relation to the steady state error. The system type is classified as system type 1 and system type 2. When the reference command is the step function, the output settles to the reference command without error in the system type 1. But the error-integral settles not to zero but to a finite value. However, in system type 2, both error and error-integral settle to zero. In ordinary textbooks, the explanations of system type 1 and 2 are made with the ramp input function in the following manner. When the reference command is the ramp function, the speed of the output is equal to the speed of the command, but there remains some position error. In system type 2, both speed error and position error settle to zero. These explanations are equivalent to the previous explanations by error and error-integral.

The system type 1 canonical transfer function is of the order 0, and the closed-loop and open-loop transfer functions are as follows:

$$T_0(s) = \frac{a_0}{a_n s^n + \cdots + a_1 s + a_0}. \tag{2.67}$$

$$G_0(s) = \frac{a_0}{a_n s^n + \cdots + a_1 s}. \tag{2.68}$$

The system type 2 canonical transfer function is order [1, 0] and the closed-loop and open-loop transfer functions are as follows:

$$T_{1,0}(s) = \frac{a_1 s + a_0}{a_n s^n + \cdots + a_1 s + a_0}. \tag{2.69}$$

$$G_{1,0}(s) = \frac{a_1 s + a_0}{a_n s^n + \cdots + a_2 s^2}. \tag{2.70}$$

Now $T_0(s)$ will be shown to be the system type 1. The reference command y_r is related to the output y as is shown in the following equation:

$$(a_n s^n + \cdots + a_1 s + a_0) y = a_0 y_r. \tag{2.71}$$

When $y_r = 1$, the steady state solution of y is 1, and the error is zero. Then error-integral is examined. The integral of y_r is represented by y_{rI}, and the integral of y is represented by y_I. Then the equation similar to Eq. (2.71) is derived.

$$(a_n s^n + \cdots + a_1 s + a_0) y_I = a_0 y_{rI}. \tag{2.72}$$

When $y_r = 1$, $y_{rI} = t$. Then the steady state of y_I is $t - a_1/a_0$, as is confirmed by direct substitution in Eq. (2.72). The error-integral takes a finite value as follows:

$$[y_{rI} - y_I]_{steady\ state} = t - (t - a_1/a_0) = a_1/a_0. \tag{2.73}$$

Thus, $T_0(s)$ is certainly the system type 1. Now the explanation is given when the reference command is the ramp function. Then $y_r = t$ in Eq. (2.71). The steady state value of y is $t - a_1/a_0$ as is clear by the direct substitution in Eq. (2.71). The speed is the same as the reference, but position has the finite error.

$$[y_r - y]_{steady\ state} = t - (t - a_1/a_0) = a_1/a_0. \tag{2.74}$$

The relation of error-integral to the step input is the same as that of the error to the ramp input.

Next, $T_{1,0}(s)$ is shown to be the system type 2. The reference command y_r is related to the output y in the following relation:

$$(a_n s^n + \cdots + a_1 s + a_0)y = (a_1 s + a_0)y_r. \tag{2.75}$$

For $y_r = 1$, the steady state value of y is 1, and the error is 0. For the error-integral, the following relation is used as Eq. (2.72).

$$(a_n s^n + \cdots + a_1 s + a_0)y_I = (a_1 s + a_0)y_{rI}. \tag{2.76}$$

For $y_r = 1$, $y_{rI} = t$. The steady state value of y_I is confirmed to be t by the direct substitution in Eq. (2.76). Because the error-integral is also 0, $T_{1,0}(s)$ is certainly the system type 2. Now the explanation is given when the reference command is the ramp function. Then $y_r = t$ in Eq. (2.75). The steady state vale of y is t as is clear by the direct substitution in Eq. (2.75). The speed and position have no error.

Finally, **CDM-type Bode Diagram** will be explained. In CDM, the design proceeds with the characteristic polynomial as the central issue with the aid of the coefficient diagram. In the classical control, on the other hand, the design proceeds with the open-loop transfer function as the central issue with the aid of the Bode diagram. When the relation of both approaches becomes clear, the design by CDM will be interpreted in terms of classical control. Suppose that the characteristic polynomial is given. If the system type is specified, the order of the canonical transfer function is obtained, and the corresponding canonical open-loop transfer function is uniquely obtained. When the open-loop transfer function is expressed on the Bode diagram, it is equivalent to the coefficient diagram of the characteristic polynomial. In order to keep close relation between the coefficient diagram and the Bode diagram, the CDM-type Bode diagram is introduced, which is the modification of the asymptotic Bode diagram.

First, the ith-order break point ω_i is defined as follows:

$$\omega_i = a_i/a_{i+1} = 1/\tau_i. \tag{2.77}$$

The ith order break point is the reciprocal of the ith order equivalent time constant τ_i, and the ratio of adjacent break points is the stability index γ_i.

$$\gamma_i = \omega_i/\omega_{i-1}. \tag{2.78}$$

Next, the definition of the CDM-type Bode diagram is given in the following: The transfer function is given as follows:

$$F(s) = \frac{b_m s^m + b_{m-1} s^{m-1} + \cdots + b_1 s + b_0}{a_n s^n + a_{n-1} s^{n-1} + \cdots + a_1 s + a_0}. \tag{2.79}$$

The CDM-type Bode diagram $F(\omega)_{CDM}$ is defined as follows:

$$F(\omega)_{CDM} = \frac{\max_j(|b_j \omega^j|)}{\max_i(|a_i \omega^i|)}. \tag{2.80}$$

The frequency response is obtained by the ratio of the approximate numerator/denominator frequency responses. The approximate denominator/numerator frequency responses can be obtained by selecting the term which has the largest absolute value in the denominator/numerator when s is set to $j\omega$. The result is similar to the asymptotic Bode diagram with the difference that the break points are no longer poles and zeros. Each pole is replaced by the ith order break point ω_i, and each zero is replaced by the break point of the numerator polynomial.

Figure 2.11 shows the CDM-type Bode diagram of system type 1 and 2 canonical open-loop transfer functions. The characteristic polynomial $P(s)$ and system type 1 and 2 canonical open-loop transfer function $G_0(s)$ and $G_{1,0}(s)$ are shown as follows:

$$P(s) = 0.015625s^5 + 0.125s^4 + 0.5s^3 + s^2 + s + 0.4. \tag{2.81}$$

$$G_0(s) = \frac{0.4}{0.015625s^5 + 0.125s^4 + 0.5s^3 + s^2 + s}. \tag{2.82}$$

$$G_{1,0}(s) = \frac{s + 0.4}{0.015625s^5 + 0.125s^4 + 0.5s^3 + s^2}. \tag{2.83}$$

Because this diagram has the one to one correspondence with the coefficient diagram, it can be easily constructed from the coefficient diagram. When the design result of CDM is expressed in this diagram, the meaning of the CDM design will be easily

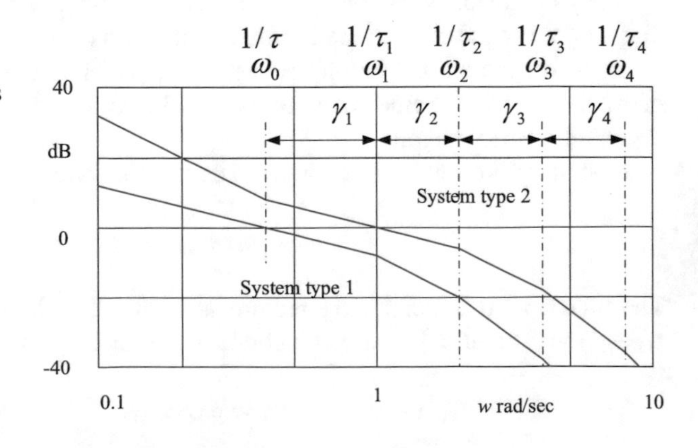

Fig. 2.11 CDM-type Bode diagram for canonical open-loop transfer functions

understood from the point of view of the classical control. The CDM-type Bode diagram has the similar degree of approximation as the asymptotic Bode diagram. At the region near the break points, the CDM-type Bode diagram shows smaller value than the asymptotic Bode diagram. The exact frequency response lies between the two.

Naslin already named the ith order break point ω_i as "characteristic pulsatance" [5]. Because the term means "characteristic frequency" and the meaning is ambiguous, the new name, the i-order break point, is introduced.

2.5 Standard Form

In this section, the standard form used in CDM will be shown. Such standard form retains many favorable characteristics as a good control system. The standard form has emerged from many design experiences, and it retains various favorable characteristics with a good balance. First **CDM Standard Form** is shown, and its **Robustness** is examined. Then it is shown that the **Step Responses** are favorable shapes and the **Pole Location** is compatible with the one recommended in the control system design. Finally, the **Features of CDM Standard Form** are summarized, and the comparison is made with other **Various Standard Forms** recommended so far in the control system design.

CDM Standard Form is the center of CDM, and is based on the following stability indices:

$$\gamma_{n-1} = \cdots = \gamma_2 = 2, \quad \gamma_1 = 2.5. \tag{2.84}$$

The background that such stability indices are being used in the control system design is as follows; Kessler started to use the stability index in the control system design and proposed the so-called Kessler standard form, where all stability indices are specified as 2 [6]. This value was found by experience through the design of steel mill drive control. However, when viewed from the present knowledge and experience, this design equivalently guarantees "the proper stability" and "the proper robustness for the variation of each coefficient of the characteristic polynomial". Naslin used the stability index in the name of characteristic ratio, but he used it more or less as the adjustment parameters to attain the design objectives such as the proper response shape, contrary to Kessler who considered the stability index as the design goal. In the design example, Naslin used the stability index of 1.75 through 2.4 [5]. The shortcoming of the Kessler standard form is that the step response, or in more exact terms the step response of the 0th order canonical transfer function, has the overshoot of about 8%. This shortcoming has been later overcome by increasing the 1st order stability index to $\gamma_1 = 2.5$, and this improvement is incorporated in the CDM standard form.

Because the CDM design proceeds with the coefficient diagram as the graphical expression of the characteristic polynomial, it is convenient to choose the standard form of the characteristic polynomial in such a manner that its expression in the coefficient diagram is natural and easy. The coefficients of the characteristic polynomial are expressed by the stability indices, the equivalent time constant, and the 0th order coefficient as in Eq. (2.7). It is shown again for reference as follows:

$$a_i = a_0\tau^i/(\gamma_{i-1}\gamma_{i-2}^2\cdots\gamma_2^{i-2}\gamma_1^{i-1}). \tag{2.85}$$

When $a_0 = 0.4$ and $\tau = 2.5$ are chosen, the characteristic polynomial $P(s)$ of the CDM standard form becomes as follows:

$$\begin{aligned} P(s) = 2^{-\frac{(n-2)(n-1)}{2}}s^n + \cdots + 2^{-10}s^6 + 2^{-6}s^5 \\ +2^{-3}s^4 + 0.5s^3 + s^2 + s + 0.4. \end{aligned} \tag{2.86}$$

The coefficient diagram of this standard form is easy to handle, because the middle portion of the diagram goes up and both ends go down, and the total shape is round (convex). This standard form is easy to remember because the ratio of the adjacent coefficients is very systematic as shown in the following:

$$a_{i+1}/a_i = 2^{(1-i)}, \qquad i = 1\cdots n-1. \tag{2.87}$$

In other standard forms currently used in control system design, it is common to choose $a_0 = 1$ and $\tau = 1$. For this choice, a different CDM standard form is obtained.

$$\begin{aligned} P(s) = 0.1^{n-1}2^{\frac{(6-n)(n-1)}{2}}s^n + \cdots + 0.00001s^6 + 0.0004s^5 \\ +0.008s^4 + 0.08s^3 + 0.4s^2 + s + 1. \end{aligned} \tag{2.88}$$

When this standard form is expressed in the coefficient diagram, the shape becomes left-end-down, and more or less straight rather than round. Although such diagram is not easy to handle, the coefficients are easy to remember. The ratio of the adjacent coefficients becomes as follows:

$$a_{i+1}/a_i = 0.1 \times 2^{(3-i)}, \qquad i = 1\cdots n-1. \tag{2.89}$$

The coefficient diagram is easy to handle when it is a round shape with the middle portion up. When it is too much left-end-down, it is recommended to draw the coefficient diagram for $a_i \times 10^i$ or $a_i \times 100^i$ instead of a_i. For this reason, the CDM standard form is chosen as the one for $a_0 = 0.4$ and $\tau = 2.5$.

Now the **Robustness** of the standard form will be examined. For Kessler's standard form, the system is stable for the variation of each coefficient for about 0.5 through 3 times. This can be explained by the law from experience that the stability condition of the partial fourth-order polynomial can be used for stability test of the total system instead of the sufficient condition for stability by Lipatov. When some coefficient of the characteristic polynomial is decreased, the worst-case partial fourth-order polynomial is the polynomial whose center term corresponds to that coefficient.

When some coefficient of the characteristic polynomial is increased, the worst-case partial fourth-order polynomial is the polynomial whose end term corresponds to that coefficient. The partial fourth-order polynomial is shown as follows:

$$P_i(s) = a_{i+4}s^4 + a_{i+3}s^3 + a_{i+2}s^2 + a_{i+1}s + a_i. \quad (2.90)$$

The stability condition for the fourth-order polynomial is given in Eq. (2.51). The stability condition of the partial fourth-order polynomial becomes as follows:

$$a_{i+2} > a_{i+4}(a_{i+1}/a_{i+3}) + a_i(a_{i+3}/a_{i+1}). \quad (2.91)$$

In Kessler's standard form, all stability indices are 2. After proper transformation of a_0 and τ, the partial fourth-order polynomial becomes as follows:

$$P_i(s) = 0.5s^4 + 2^{0.5}s^3 + 2s^2 + 2^{0.5}s + 0.5. \quad (2.92)$$

The stability condition becomes as follows:

$$2 > 0.5(2^{0.5}/2^{0.5}) + 0.5(2^{0.5}/2^{0.5}) = 0.5 + 0.5. \quad (2.93)$$

From this equation, it is clear that the system becomes unstable when a_{i+2} is decreased to one half, or when a_{i+4} is increased to 3 times. The above is the explanation that the system is stable for the variation of each coefficient for about 0.5 through 3 times in Kessler's standard form. The above is an approximation, and there is some error. But such error can be neglected in practice. The robustness for the coefficients of both ends is different from the above general case. The a_n and a_0 can be decreased to 0 times, and a_{n-1} and a_1 can be decreased to 0.26 times. All of them can be increased to 3 times, the same as the general case. In CDM standard form, a_0 is smaller by the ratio of γ_1. $2/2.5 = 0.8$. The allowable increase becomes 4 times. The limit of decrease for a_1 is about 0.2 times. For $\gamma_i = 1.75$, as used by Naslin, the allowable variation is about 0.65 through 2 times, narrower than Kessler.

More generally, it is possible to determine the Hurwitz stability of a ball of polynomials specified by a weighted l_p norm in the coefficient space for an arbitrary positive integer p [7, see Sect. 3.4]. Tsypkin and Polyak locus is a graphical approach for this problem. In order to examine the robustness of a family of interval polynomials with CDM standard form, the maximal radius of the l_∞ stability ball will be determined and compared. Now, consider a family of polynomials $P(s)$ centered at a nominal point $\mathbf{a}_i^o$ with coefficients lying in the weighted l_∞ ball $\mathcal{B}_\infty(\mathbf{a}_i^o, \rho)$ of radius ρ

$$P(s) = a_n s^n + \cdots + a_1 s + a_0, \quad (2.94)$$

$$\mathcal{B}_\infty(\mathbf{a}_i^o, \rho) := \left\{ \mathbf{a}_i \left[max_k \left| \frac{a_k - a_k^o}{\alpha_k} \right| \right] \le \rho \right\}, \quad (2.95)$$

where $\mathbf{a}_i = [a_n, \;\cdots,\; a_1\; a_0]$, $\alpha_k > 0$ are given weights, and $\rho \ge 0$ is a prescribed common margin for the perturbations. Note that this set $P(s)$ is a family of polynomials with interval uncertainty. If we set $\alpha_k = a_k^o$, ρ indicates the percentage variation of each coefficient at the nominal parameters vector. That is, every polynomial with $\mathbf{a}_i$ in $\mathcal{B}_\infty(\mathbf{a}_i^o, \rho)$ retains stability. Thus, polynomials that give larger ρ are more robust.

As an example, let us first find the l_∞ stability bound of the fourth-order polynomial with Kessler form (*i.e.*, $\gamma_i = [2\ 2\ 2]$). When $\tau = 1$, the nominal coefficient vector of the polynomial is given by

$$\mathbf{a}_i^o = [a_n^o, \ \cdots, \ a_1^o \ a_0^o] = [1 \ \ 8 \ \ 32 \ \ 64 \ \ 64].$$

We choose $\alpha_k = a_k^o$ to obtain the percentage variations of coefficients. Then using Tsypkin and Polyak's theorem [7, Theorem 3.7], the maximum ball radius that guarantees robust stability is $\rho = 0.2679$ (26.79%).

Now, we find the l_∞ stability bound of the fourth-order polynomial with CDM standard form (*i.e.*, $\gamma_i = [2\ 2\ 2.5]$) when $\tau = 1$. The nominal coefficient vector of the polynomial is given by

$$\mathbf{a}_i^o = [a_n^o, \ \cdots, \ a_1^o \ a_0^o] = [1 \ \ 10 \ \ 50 \ \ 125 \ \ 125].$$

the maximum ball radius that guarantees robust stability is $\rho = 0.28286$ (28.29%). In other words, a family of 4th-order polynomials with CDM standard form retains stability if their coefficients are in the following intervals:

$$a_4 \in [0.7172, \ 1.28286], \ \ a_3 \in [7.175, \ 12.8286], \ \ a_2 \in [35.857, \ 64.143], \\ a_1 \in [89.643, \ 160.3575], \ a_0 \in [89.643, \ 160.3575].$$

The stability tolerance of the CDM polynomial is 1.5% larger than that of the Kessler polynomial, thus it reveals that the latter is more robust than the former. In a similar way, it can be seen that increasing each γ_i increases the robustness.

The **Step Response** of the standard form will be discussed hereafter. For the CDM standard form with $a_0 = 0.4$ and $\tau = 2.5$, the step responses of the system type 1 canonical transfer function $T_0(s)$ of various orders are shown in Fig. 2.12. The overshoot is virtually zero, and the responses are about the same, irrespective of the order of the system. Because of this nature, the designer can start from a simple controller and move smoothly to a more complicated one in addition to the previous design. The settling time is about 2.5–3 times the equivalent time constant τ. It is found by simulation that the settling time of the CDM standard form looks shortest compared with other standard forms with the same equivalent time constant. Here the term "settling time" is not exactly defined; rather commonly accepted notion, when engineers see the response, is adopted.

Step responses of the system type 2 canonical transfer function, $T_{1,0}(s)$, of various orders are shown in Fig. 2.13. There is an overshoot of about 40% for system type 2. This overshoot is necessary because the integral of the error, as well as the error itself, must become zero in system type 2. The responses are about the same, irrespective

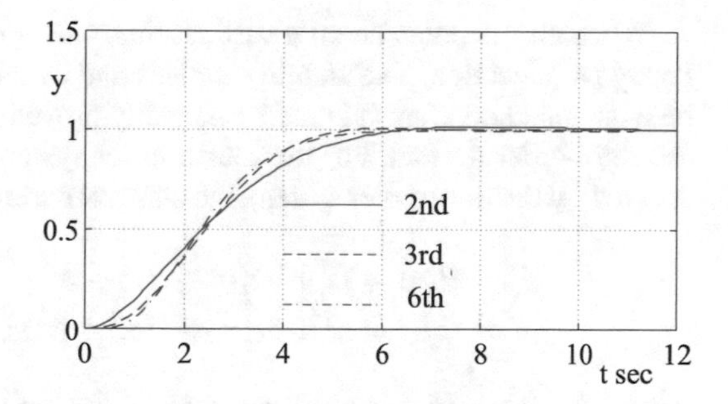

Fig. 2.12 Step responses of the type 1 canonical form $T_0(s)$ of degree 2, 3, 6

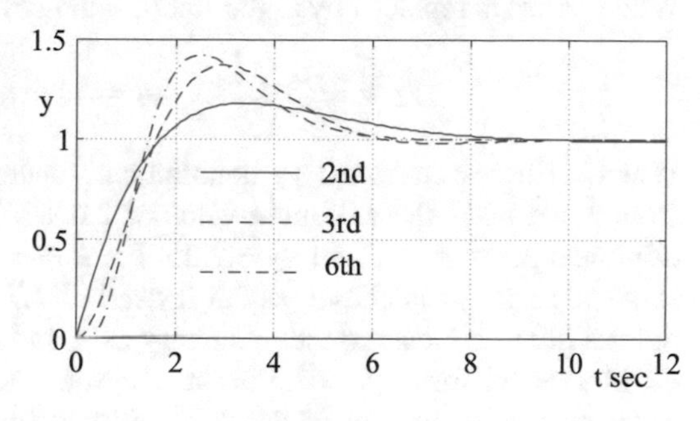

Fig. 2.13 Step responses of the type 2 canonical form $T_1(s)$ of degree 2, 3, 6

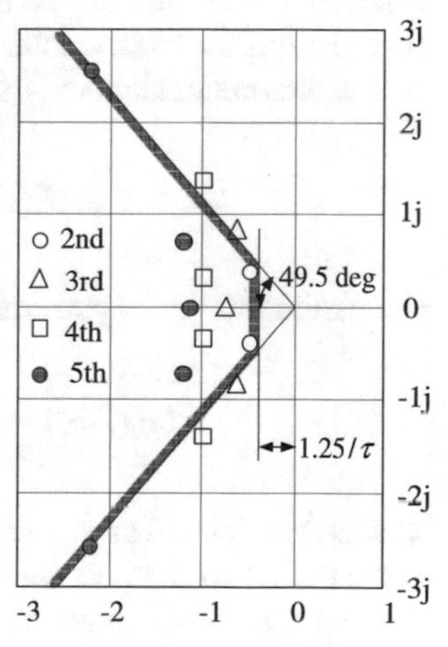

Fig. 2.14 Pole location

of the order of the system above the third order. The settling time is about 2.5–3τ, the same as the system type 1.

Pole Location will be discussed next. The pole location of CDM standard form is shown in Fig. 2.14. It is found that the three lowest order poles are aligned in a vertical line and the two highest order poles are at the point about 49.5° from the negative real axis. The rest of the poles are on or close to the negative real axis in the region surrounded by such vertical lines and sectors. For the fourth order, all poles are exactly on the vertical line. It is well known that, when poles lie in the region surrounded by a vertical line and a sector, the system shows a good response.

When all the poles lie on a vertical line, such pole location will be called equal decay pole location. The stability indices and overshoot for such pole location will be examined hereafter. It is mathematically proven [8] that there is no overshoot at the third-order system. For the fourth-order system, the no overshoot condition is derived. At the third-order system, the characteristic polynomial becomes as follows:

$$\begin{aligned} P(s) &= (s+\alpha)[(s+\alpha)^2+\beta^2], \\ &= s^3+3\alpha s^2+(3\alpha^2+\beta^2)s+\alpha(\alpha^2+\beta^2). \end{aligned} \tag{2.96}$$

When β^2/α^2 is replaced by k, the stability indices becomes as follows:

$$\gamma_2 = 9/(3+k), \quad \gamma_1 = (3+k)^2/[3(1+k)]. \tag{2.97}$$

When k is increased from 0, γ_2 monotonically decreases from 3 to 0. But γ_1 decreases from 3 and takes the minimum value of 2.6667. Then it starts to increase. At the minimum point, $k = 1$ and $\gamma_2 = 2.25$. For $k = 1.5$, $\gamma_2 = 2$ and $\gamma_1 = 2.7$. In CDM standard form, γ_1 is chosen as 2.5 instead of 2.7. This choice moves the complex poles a little bit closer to the imaginary axis, and a small overshoot appears. In the CDM standard form, $\gamma_1 = 2.5$ is chosen even with this small overshoot at the third order, simply because, by so doing, stability indices can be the same, irrespective of the order of the characteristic polynomial, and the numbers are easy to remember.

The 0th order canonical transfer function for this system is as follows:

$$T_0(s) = \frac{\alpha(\alpha^2+\beta^2)}{(s+\alpha)[(s+\alpha)^2+\beta^2]}. \tag{2.98}$$

The partial fraction expansion gives

$$T_0(s) = \left[\frac{1}{s+\alpha} - \frac{s+\alpha}{(s+\alpha)^2+\beta^2}\right]\frac{\alpha(\alpha^2+\beta^2)}{\beta^2}. \tag{2.99}$$

Now $z(t)$ is defined as the output to the unit impulse input, $\delta(t)$, and $y(t)$ is defined as the output to the unit step input, $u_s(t)$. Naturally, $z(t)$ is the derivative of $y(t)$. Here $z(t)$ can be obtained by inverse Laplace transform of Eq. (2.99) [3, 4]. In the similar manner, $y(t)$ can be obtained from $T_0(s)/s$. The results are as follows:

$$z(t) = e^{-\alpha t}[1-\cos\beta t]\alpha(\alpha^2+\beta^2)/\beta^2. \tag{2.100}$$

$$y(t) = 1 - e^{-\alpha t}[(\alpha^2+\beta^2)/\beta^2 - (\alpha^2/\beta^2)\cos\beta t + (\alpha/\beta)\sin\beta t)]. \tag{2.101}$$

From Eq. (2.100), it is clear that $z(t)$ is non-negative. Thus, $y(t)$ monotonically increases and there is no overshoot.

A similar analysis will be made for the fourth-order system. The characteristic polynomial for the fourth-order system is given as follows:

$$\begin{aligned} P(s) &= [(s+\alpha)^2 + \beta_1^2][(s+\alpha)^2 + \beta_2^2], \\ &= (s+\alpha)^4 + (\beta_1^2 + \beta_2^2)(s+\alpha)^2 + \beta_1^2\beta_2^2, \\ &= s^4 + 4\alpha s^3 + (6+k_1)\alpha^2 s^2 + (4+2k_1)\alpha^3 s + (1+k_1+k_2)\alpha^4. \end{aligned} \tag{2.102}$$

In the above equations, k_1 and k_2 are defined as follows:

$$(\beta_1^2 + \beta_2^2)/\alpha^2 = k_1, \quad \beta_1^2\beta_2^2/\alpha^4 = k_2. \tag{2.103}$$

For $k_1 = 2$, the stability indices becomes as follows:

$$\gamma_3 = \gamma_2 = 2, \quad \gamma_1 = 8/(3+k_2). \tag{2.104}$$

Because $k_1 = 2$, it is clear from Eq. (2.103) that the value of k_2 is limited within 0 and 1. Thus γ_1 is limited within 2.6667 through 2 by Eq. (2.104). For $k_2 = 0$, $\gamma_1 = 2.6667$. For $k_2 = 0.2$, $\gamma_1 = 2.5$ and the system becomes CDM standard form. For $k_2 = 1$, $\gamma_1 = 2$ and the system becomes Kessler's standard form.

The 0th order canonical transfer function for this system is as follows:

$$T_0(s) = \frac{(\alpha^2 + \beta_1^2)(\alpha^2 + \beta_2^2)}{[(s+\alpha)^2 + \beta_1^2][(s+\alpha)^2 + \beta_2^2]}. \tag{2.105}$$

By the similar treatment as in the third-order system, next results are obtained.

$$z(t) = te^{-\alpha t}\left[\frac{\sin\beta_2 t}{\beta_2 t} - \frac{\sin\beta_1 t}{\beta_1 t}\right]\frac{(\alpha^2 + \beta_1^2)(\alpha^2 + \beta_2^2)}{\beta_1^2 - \beta_2^2}. \tag{2.106}$$

$$\begin{aligned} y(t) = 1 &- e^{-\alpha t}\frac{\alpha^2 + \beta_1^2}{\beta_2(\beta_1^2 - \beta_2^2)}(\beta_2\cos\beta_2 t + \alpha\sin\beta_2 t) \\ &- e^{-\alpha t}\frac{\alpha^2 + \beta_2^2}{\beta_1(\beta_1^2 - \beta_2^2)}(\beta_1\cos\beta_1 t + \alpha\sin\beta_1 t). \end{aligned} \tag{2.107}$$

From Eq. (2.106), it is clear that $z(t)$ is non-negative, if $\beta_2 \ll \beta_1$, and there is no overshoot in $y(t)$, when $k_1 = 2$, and α is large. For $\beta_2 = 0$, $k_2 = 0$ and $\gamma_1 = 2.6667$; there is no overshoot. For $\beta_2 = 0.23607\beta_1$, $k_2 = 0.2$ and $\gamma_1 = 2.5$; this is CDM standard form, and there is no overshoot. For $\beta_2 = \beta_1$, $k_2 = 1$ and $\gamma_1 = 2$; this is Kessler's standard form and there is some overshoot. Different from the third-order system, the equal decay pole location does not guarantee no overshoot in the fourth-order system.

By the exact calculation, the overshoot in Kessler's standard form is 8.15% in the third-order system, and 6.24% in the fourth-order system. For the fifth order or above, the overshoot is the same as the fourth order. However, the overshoot of Kessler's standard form is usually called 8% on the basis of the third-order system.

The **Features of CDM Standard Form** are summarized hereafter. The CDM standard form has the following favorable characteristics:

(1) The system has strong robustness to the variation of each coefficient of the characteristic polynomial.
(2) For system type 1, overshoot is almost zero. For system type 2, a necessary overshoot of about 40% is realized.
(3) Among the systems with the same equivalent time constant τ, the CDM standard form seems to have the shortest settling time. The settling time is about 2.5–3τ.
(4) The step responses show almost equal waveforms irrespective of the order of the characteristic polynomial.
(5) The lower order poles are aligned on a vertical line. The higher order poles are located within a sector 49.5° from the negative real axis, and their damping ratio ζ is larger than 0.65.
(6) The CDM standard form is very easy to remember.

In other words, the CDM standard form seems to possess all the features of "Good control system" found from experience, such as "Strong robustness", "No overshoot", "Short settling time", and "Poles are located in the region surrounded by a vertical line and a sector".

Comparison of **Various Standard Forms** will be made hereafter. These standard forms are recommended from various standpoints in the past. The stability indices of these standard forms are listed in Table 2.1. In binomial, Butterworth, Bessel, and ITAE (integral of time multiplied by absolute error), the stability indices vary with the order of the respective characteristic polynomials. However, they are constant in Kessler, CDM, and Kitamori.

The stability indices of the binomial standard form are obtained from the coefficients of the characteristics polynomial, which are ${}_iC_n$ when the polynomial is normalized as $a_n = a_0 = 1$. The stability indices are shown in the following formula:

$$\gamma_i = \left(\frac{i+1}{i}\right)\left(\frac{n-i+1}{n-i}\right). \tag{2.108}$$

The stability indices of Bessel's standard form are obtained from the coefficients a_i of the characteristic polynomial, which are given for the normalized form of $a_n = 1$ and $\tau = 1$ [9]. The stability indices are shown in the following formula:

$$\gamma_i = \left(\frac{i+1}{i}\right)\left(\frac{n-i+1}{n-i}\right)\left(\frac{2n-i}{2n-i+1}\right). \tag{2.109}$$

The stability indices of Butterworth's standard form are calculated from the ratios of the adjacent coefficients a_{i+1}/a_i, which are given for the normalized form of $a_n = a_0 = 1$ [9]. The stability indices are shown in the following formula:

$$\gamma_i = 1 + \frac{\sin(\pi/n)}{\sin(i\pi/n)}. \tag{2.110}$$

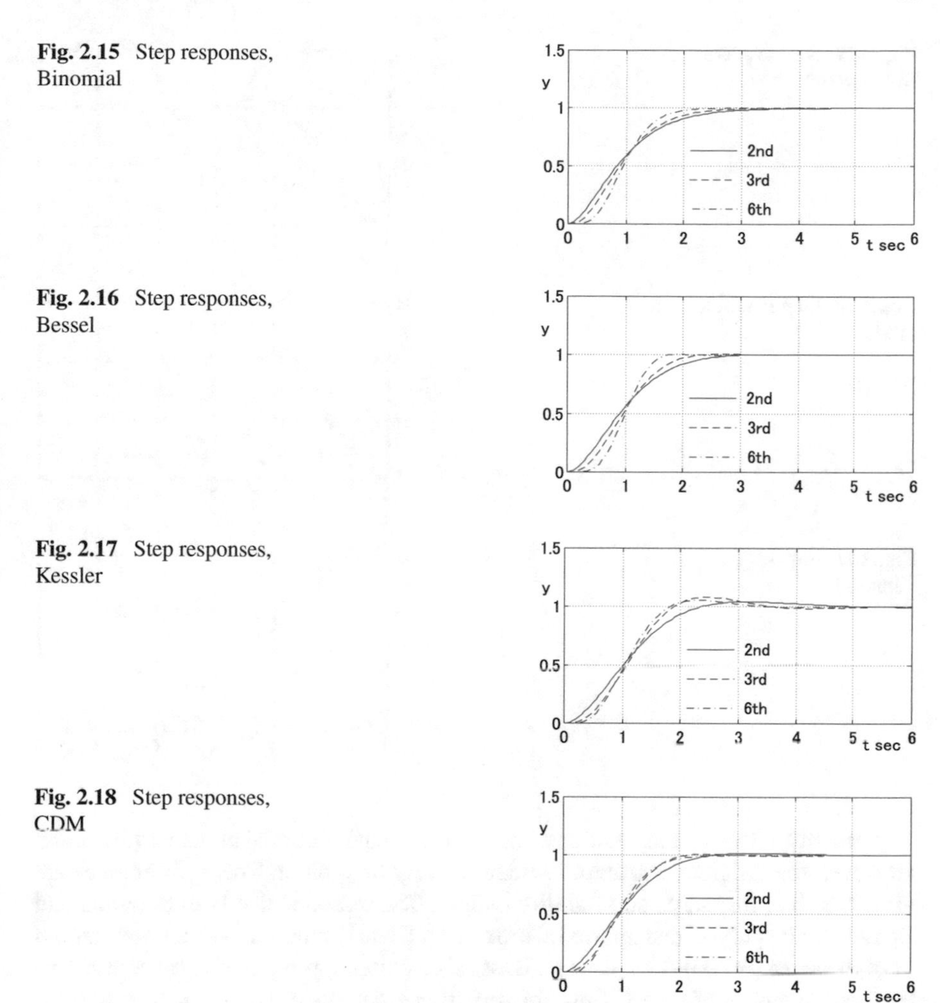

Fig. 2.15 Step responses, Binomial

Fig. 2.16 Step responses, Bessel

Fig. 2.17 Step responses, Kessler

Fig. 2.18 Step responses, CDM

The stability indices of ITAE are calculated from the coefficient a_i, which were given for the normalized form of $a_n = a_0 = 1$ [10]. The stability indices of Kitamori are calculated from the coefficients a_i, which are given for the normalized form of $a_n = 1, \tau = 1$ [11].

The step responses of the system type 1 canonical transfer function $T_0(s)$ of various orders and of various standard forms are given in Figs. 2.15, 2.16, 2.17, 2.18, 2.19, 2.20 and 2.21.

From these Table and Figs., the following conclusions are derived. Firstly, the overshoot is greatly affected by the value of the stability index γ_1. According to the sensitivity analysis of Kim et al. [12] about how much individual stability index affects the transient response of system $T_0(s)$, the low-order stability indices are dominant than the high-order ones. If it is larger than 2 with the appropriate margin, no overshoot is realized, as evidenced from binomial, Bessel, and CDM.

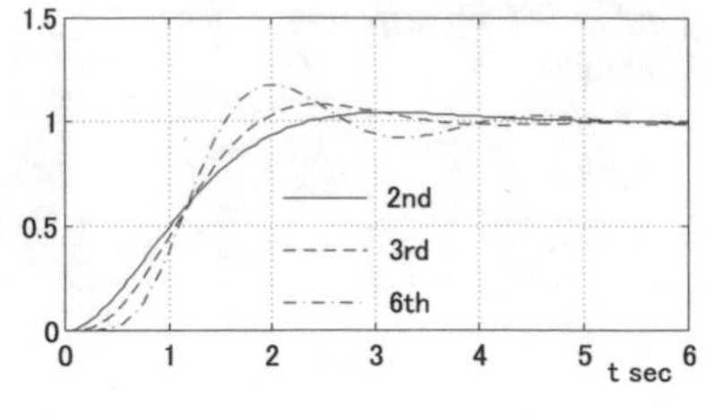

Fig. 2.19 Step responses, Butterworth

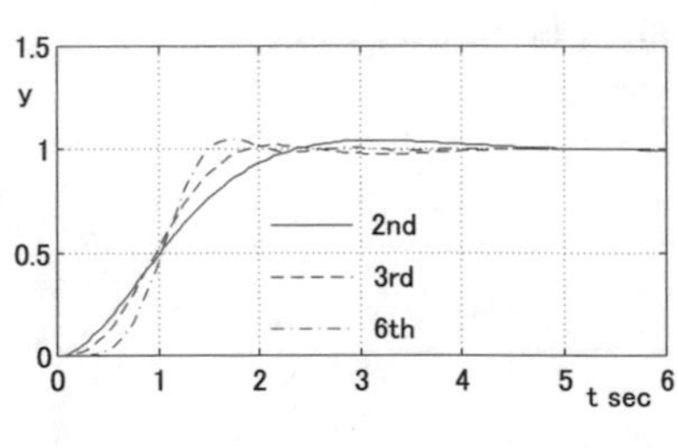

Fig. 2.20 Step responses, ITAE

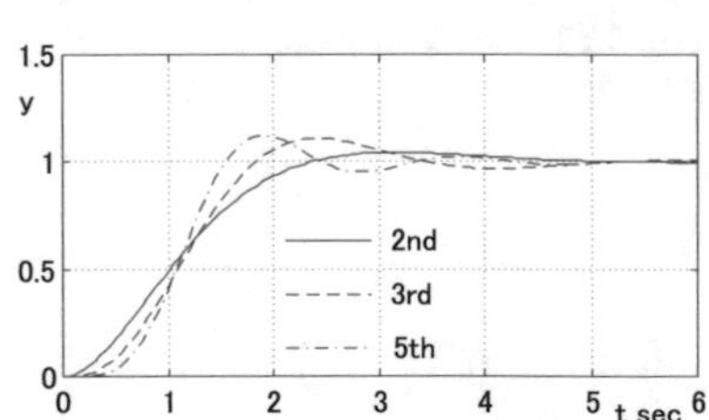

Fig. 2.21 Step responses, Kitamori

Secondly, CDM resembles Bessel at low order, and binomial at the middle order. However, the differences become evident at very high order. These differences are visible both in responses and stability indices. The reason is that both binomial and Bessel are the polynomial approximation of e^s. Clearly binomial is a approximation of e^s, because $[(s/n+1)^{n/s}]^s \sim e^s$. Bessel is a better approximation by continuous fraction expansion of e^s [9]. Thus, for very high order, the responses of both systems become the time delay response of the equivalent time constant $\tau = 1$ with inherent steep rise. The settling time approaches $\tau = 1$. From Eqs. (2.108) and (2.109), it is clear that the stability indices of binomial at high orders are smaller than those of CDM, and those of Bessel are further smaller. In binomial and Bessel at high orders, the stability indices of at both ends, γ_{n-1} and γ_1, approach to 2, while the stability indices at the middle γ_i, $i \simeq n/2$, approach to 1. In CDM, on the other hand, both stability indices and response waveforms do not change by the increase of the order. It does not show the time delay response like high-order binomial or Bessel, and the discrepancy becomes conspicuous.

Thirdly, the settling time of CDM is the smallest in a practical sense. In a very high-order binomial or Bessel, the settling time becomes equal to the equivalent time constant $\tau = 1$, shorter than that of CDM. However, this is achieved by smaller stability indices, hence in the sacrifice of robustness. These cases have to be excluded in a practical sense.

Table 2.1 Comparison of stability index

Standard forms	Stability indices				
	γ_5	γ_4	γ_3	γ_2	γ_1
Binomial					4
				3	3
			2.6667	2.25	2.6667
		2.5	2	2	2.5
	2.4	1.875	1.7778	1.875	2.4
Bessel					3
				2.4	2.5
			2.2222	1.9286	2.3333
		2.1429	1.75	1.7778	2.25
	2.1	1.6667	1.6	1.7045	2.2
Butterworth					2
				2	2
			2	1.7071	2
		2	1.6180	1.6180	2
	2	1.5774	1.5	1.5774	2
ITAE					2
				1.4244	2.6414
			1.2971	2.0388	2.1441
		1.568	1.6234	1.7794	2.1018
	1.6004	1.5585	1.5042	1.6339	2.0943
Kitamori		2	1.5	1.6667	2
Kessler	2	2	2	2	2
CDM	2	2	2	2	2.5

2.6 Robustness Consideration

As shown in the previous section, the standard form in CDM has many favorable characteristics as a good control system. Especially it should be noted that the standard form guarantees robustness as well as stability. But under special circumstances, the stability and robustness become related in trade-off condition, and both cannot be satisfied at the same time. Under such situations, it is not advisable to stick to the CDM standard form, and some kinds of modifications are necessary. The purpose of this section is to give some guidelines for such modification.

Firstly, the meanings of **Stability and Robustness** as used in this book will be explained. Secondly, as an example of a stable but weakly robust system, **Multiple Pole System** will be discussed. Thirdly, as another example of unrobust controller, **Unstable Poles and Non-minimum-phase Zero** controllers are discussed. Several examples are shown to clarify the inherent problems. Finally, specific recommendations are proposed for **Modification of CDM Standard Form**.

Stability and Robustness are considered as a similar concept in ordinary control systems, the system with good stability is usually robust. In the classical control, such indices as the phase margin and gain margin are the indices for stability as well as robustness. Simply stated, stability concerns how far the poles are located to the left of the imaginary axes in the complex plane, while robustness concerns how quickly poles move to the imaginary axes for the given parameter variation of the plant. If the speed of pole movement for the parameter variation is the same, larger parameter variation is necessary for the poles to move to the imaginary axes in a very stable system, because the poles are located far to the left side. Thus, the stable system is also robust.

However, if the pole moves sensitively for the small variation of parameters, robustness becomes low even with sufficient stability. One such example is the case of multiple poles. In the CDM standard form, stability, as well as robustness, is high, because poles are distributed over a wide surface. If the poles are concentrated, as in the multiple poles, the stability indices become generally small. Also, the poles move sensitively to the variation of parameters, and robustness becomes low even if the system is sufficiently stable. Another example is the case where the controller has unstable poles and non-minimum-phase zeros. Such controller becomes necessary in control of the plants where control is very difficult. In such systems, the coefficients of the characteristic polynomial become the difference of two parameters and vary extensively to the small variation of parameters. For this reason, robustness becomes low even for the characteristic polynomial in the CDM standard form. In such a situation, it is advisable to reduce the stability indices to exclude unstable poles and non-minimum-phase zero from the controller as far as possible and attain high robustness. The stability and robustness become the trade-off issue and both cannot be satisfied at the same time. The stability must be sacrificed for the robustness to some extent.

Now the case of **Multiple Pole System** will be considered. In order to make the analysis simple, the binomial standard form is considered. The characteristic polynomial is expressed as follows:

$$P(s) = (s+1)^n = a_n s^n + a_{n-1} s^{n-1} + \cdots + a_1 s + a_0 = \sum_{i=0}^{n} a_i s^i,$$

$$a_i = {}_iC_n = \frac{n!}{i!(n-i)!}. \tag{2.111}$$

The equivalent time constant τ of this system is n, and its 0th order canonical transfer function converges to the time delay element e^{-ns} of delay time n for large n. Its step response has no overshoot, and it is considered very stable because poles are multiple poles at $s = -1$. Now let us make $n = 2m$, and consider the robustness when a_m varies to $(1-\delta)a_m$. The characteristic roots or poles are calculated as follows:

$$(s+1)^{2m} - \delta a_m s^m = 0. \tag{2.112}$$

It can be verified by simple calculation that a root becomes $s = j$, when $\delta = 2^m/a_m$. This δ gives the stability limit. By the Stirling's formula $n! = \sqrt{2\pi} n^{n+1/2} e^{-n}$, δ is approximated as follows:

$$\delta = 2^m/a_m = 2^m (m!)^2/(2m)! \sim \sqrt{\pi/2}\sqrt{n}\, 2^{-0.5n} = 1.2533\sqrt{n}\, 2^{-0.5n}. \tag{2.113}$$

For $n = 10$, $\delta = 0.12385$. For $n = 20$, $\delta = 0.0054736$. The robustness is very poor. As generally known, the multiple poles moves sensitively to the variation of parameters, and the robustness is low even when the stability is high. As explained in the previous section, the stability indices of binomial standard form are $\gamma_i = [(i+1)/i][(n-i+1)/(n-1)]$. The stability index γ_m, corresponding to a_m, is given as follows:

$$\gamma_m = \left(\frac{m+1}{m}\right)^2. \tag{2.114}$$

For $n = 10$ or $m = 5$, $\gamma_5 = 1.44$. For $n = 20$ or $m = 10$, $\gamma_{10} = 1.21$. This shows that the stability index becomes small in accordance with the degradation of robustness. In case of multiple poles, robustness becomes low even when stability is high, but such trouble can be predicted, if the stability indices are checked beforehand. Thus stability index is the index of stability as well as robustness.

Next, the case of the controller with **Unstable Pole and Non-minimum-phase Zero** will be explained with some examples. In ordinary systems, the parameters of denominator and numerator polynomials of plant and controller are positive. The coefficients of the characteristic polynomial are the sum of the products of such parameters. The percentage variations of coefficients are equal or smaller than the percentage variation of parameters. It has already been explained that the characteristic polynomial, designated as the CDM standard form, is highly robust against the variations in coefficients. Therefore, its robustness to the variations of the parameters of the controller/plant is also high. If some parameters of the denominator and numerator polynomials of plant and controller are negative, the coefficients of the characteristic polynomials become the sum of such parameters, and robustness becomes very low because the parameter variation is greatly amplified. Such difficulty usually takes place in control of such difficult plant as with poor controllability and/or observability, where the plant is close to unfavorable pole-zero cancellation. It may also occur when control specification is given against the natural and inherent characteristics of the plant.

The first example is the control of the oscillatory system with resonance and anti-resonance by a first-order controller. The open-loop transfer function $G(s)$ and the characteristic polynomial $P(s)$ are given as follows:

$$G(s) = G_c(s)G_p(s) = \frac{k_1 s + k_0}{l_1 s + 1}\frac{s^2+1}{s(s^2+2)}, \tag{2.115}$$

$$P(s) = l_1 s^4 + (k_1+1)s^3 + (2l_1 + k_0)s^2 + (2+k_1)s + k_0. \tag{2.116}$$

Under the condition that $k_1 \geq 0$, the controller, for which the minimum stability index is the maximum, is designed with the results that $\gamma_3 = \gamma_2 = \gamma_1 = \sqrt{3},\ k_1 = 0,\ k_0 = 1.2408,\ l_1 = 0.31020$, and $\tau = 1.6119$. For CDM standard form, where $\gamma_3 = \gamma_2 = 2, \gamma_1 = 2.5$, the designed controller is $k_1 = -070898, k_0 = 0.7691, l_1 = 0.048868$, and $\tau = 1.6786$. As in this example, non-minimum-phase zero is introduced and robustness becomes low if the CDM standard form is unduly adhered. In this system, the stability and robustness are in the trade-off relation, and both cannot be satisfied simultaneously. The design issue is to find the compromise condition, which amounts to the wise selection of k_1. The reason that such trade-off condition comes up to surface is that the ratio of squares of anti-resonance and resonance frequencies is $1 : 2$. If the ratio is $1 : 3.75$, the control becomes simple, because the CDM standard form can be realized for $k_1 = 0$. The difficulty occurs because the resonance and anti-resonance frequencies are closely located.

The second example is the control of the plant with the unstable pole and non-minimum-phase zero by a first-order controller. The open-loop transfer function $G(s)$ and the characteristic polynomial $P(s)$ are given as follows:

$$G(s) = G_c(s)G_p(s) = \frac{k_1 s - k_0}{l_1 s - 1}\frac{s - 1}{s(s - 2)}, \tag{2.117}$$

$$P(s) = l_1 s^3 + (-1 - 2l_1 + k_1)s^2 + (2 - k_1 - k_0)s + k_0. \tag{2.118}$$

In this example, the controller must have an unstable pole and a non-minimum-phase zero, and the robustness is very low. The case for the highest robustness is for $l_1 \simeq k_0 \simeq 0$ and $k_1 = 1.4142$. The system is stable for the loop gain variation of 0.70711 to 1.4142 of the nominal value. However, because $\gamma_1 \simeq \gamma_2 \simeq \infty$ and $\tau \simeq \infty$, it takes infinite time for the system to settle. Moreover, the controller is required to have an infinitely wide bandwidth corresponding to $l_1 \simeq 0$. Because of such difficulty, it is advisable to keep γ_2, γ_1, and τ within the reasonable limit. The choice of $k_1 = 1.4142$, $k_0 = 0.14645$, and $l_1 = 0.051777$ gives $\gamma_2 = \gamma_1 = 4.2426$, and $\tau = 3$. The system is stable for the loop gain variation of 0.78578 to 1.2726 of the nominal value. In this design, the robustness is sacrificed for the faster response and narrower controller bandwidth. Because the plant is difficult to control, larger stability indices are necessary compared with the CDM standard form.

The third example is the case where the plant itself is very stable. Various controllers are designed for this plant. The open-loop transfer function $G(s)$ and the characteristic polynomial $P(s)$ are given as follows:

$$G(s) = G_c(s)G_p(s) = \frac{k_1 s + k_0}{(l_3 s^3 + l_2 s^2 + s)}\frac{1}{(s + 10)}, \tag{2.119}$$

$$P(s) = (l_3 s^3 + l_2 s^2 + s)(s + 10) + k_1 s + k_0. \tag{2.120}$$

The design is made for the equivalent time constant $\tau = 1$. The simplest case is the integral control, where $l_3 = l_2 = k_1 = 0$, and $k_0 = 10$. The characteristic polynomial is as follows:

$$P(s) = s^2 + 10s + k_0 = s^2 + 10s + 10. \tag{2.121}$$

Because the stability index $\gamma_1 = 10$, the stability index is much larger than that of the CDM standard form and the robustness is high. Next design is made for $\gamma_1 = 2.5$. In this design the PI control with non-minimum-phase zero is introduced, where $l_3 = l_2 = 0$, $k_1 = -7.5$, and $k_0 = 2.5$. The characteristic polynomial is as follows:

$$P(s) = s^2 + (10 + k_1)s + k_0 = s^2 + 2.5s + 2.5. \tag{2.122}$$

Although $\gamma_1 = 2.5$ as the CDM standard form, the robustness is very low, because k_1 is negative. In the last design, the controller is chosen to be the I controller with filter, where $l_3 = 0.05$, $l_2 = 0.3$, $k_1 = 0$, and $k_0 = 10$. The characteristic polynomial is as follows:

$$P(s) = (l_3 s^3 + l_2 s^2 + s)(s + 10) + k_0 = 0.05s^4 + 0.8s^3 + 4s^2 + 10s + 10. \tag{2.123}$$

This design is close to the CDM standard form with $\gamma_3 = 3.2$, $\gamma_2 = 2$, and $\gamma_1 = 2.5$. Because this controller is stable and minimum-phase, the robustness is high.

When the plant itself is very stable, the excessive adherence to the CDM standard form tends to introduce the non-minimum-phase zero. It is advisable to avoid such a design, because the robustness becomes low. One way to avoid such a design is to use the excessively large stability index as it is. Another way is to use I control with filter to avoid non-minimum-phase zero.

Finally, **Modification of CDM Standard Form** will be discussed. As shown in the above examples, the excessive adherence to the CDM standard form may not be advisable. Some modifications of the CDM standard form may become necessary. As for the stability indices, it is strongly recommended to choose $\gamma_{n-1} = \gamma_2 = 2$, and $\gamma_1 = 2.5$. However, other stability indices can be chosen in more relaxed condition as follows:

$$\gamma_i > 1.5\gamma_i^*, \quad i = 3 \cdots n - 2. \tag{2.124}$$

This condition is equivalent to the following condition:

$$a_i > 1.5 \left(a_{i+2} \frac{a_{i-1}}{a_{i+1}} + a_{i-2} \frac{a_{i+1}}{a_{i-1}} \right), \quad i = 3 \cdots n - 2. \tag{2.125}$$

For this condition, the step response is almost the same as that of the CDM standard form. When all γ_i's are chosen to be larger than 1.5, the system is guaranteed to be stable by the sufficient condition for stability by Lipatov. Lipatov proved that, if all γ_i's are larger than 4, all roots of the characteristic polynomial are negative real [1]. Choosing γ_i larger than 4 is an excessive consideration of stability. For these reasons, it is strongly recommended that γ_i's are chosen between 1.5 and 4.

Simply stated, the stability depends on the characteristic polynomial, while the robustness is strongly affected by the structure of the controller. The traditional common sense, such as "Avoid non-minimum-phase controller as far as possible", "Use low-order controller, if possible", and "Adopt narrow bandwidth controller", seems to be a strong guarantee of robustness because such common sense tends to prevent the tendency that the coefficients of the characteristic polynomial become the difference of parameters of the plant and controller.

In the CDM design, the wise selection of stability indices γ_i is of utmost importance. Because such selection can be done only after the deep analysis of the problem from various standpoints, some experience is necessary, as is true in any design problem.

2.7 Summary

The important points in this chapter are summarized in the following. The important parameters used in CDM are defined, and their **Mathematical Relations** are clarified. The most important parameters in CDM are the stability index γ_i and the equivalent time constant τ. Also, the stability limit γ_i^* is used in the auxiliary nature. When the characteristic polynomial $P(s)$ is normalized such as $a_0 = 1$, it is expanded in the powers of (τs), where the coefficients are expressed solely as the functions of the stability indices. Thus the time scale is determined by the equivalent time constant, while the response waveform is determined by the stability indices.

At the **Coefficient Diagram**, the stability indices determine the curvature, and the equivalent time constant determines the inclination. When the stability indices are small, the stability is poor, and the robustness is also low. When the characteristic polynomial is largely separated from the component polynomials, the robustness becomes low even the stability indices are large. Thus, the stability, response, and robustness are easily read from the coefficient diagram.

The **Stability Condition**, most frequently used in CDM, is the sufficient condition for stability by Lipatov. It states that, if all the partial fourth-order polynomials of the characteristic polynomial are stable with the margin of 1.12, the total system is stable. The stability of the fourth-order polynomial can be confirmed by the check that the stability index is larger than the stability limit. It can also be confirmed by the graphical interpretation of the coefficient diagram.

In order to find the actual response, the characteristic polynomial must be converted to the transfer function by defining the appropriate numerator polynomial. For this purpose, the **Canonical Transfer Function** is defined solely by the characteristic polynomial. All the transfer functions with the given characteristic polynomial are expressed as the weighted sum of such canonical transfer functions. From the canonical transfer function, the canonical open-loop transfer function is defined. By such functions, the relation between the CDM and the classical control with the Bode diagram is clarified. At this time, the CDM-type Bode diagram is introduced, whereby the relation between the Bode diagram and the coefficient diagram becomes much closer.

In order to expedite the actual design, the CDM **Standard Form** is introduced, where all the stability indices are 2, except $\gamma_1 = 2.5$. This standard form has emerged from the long past experience. It shows good stability and high robustness. It shows a short settling time without the overshoot. The pole locations are reasonable. In ordinary control system design, the CDM standard form is the safe and effective choice.

When the plants have special characteristics, there arise some cases where some modifications of the standard form become necessary due to **Robustness Consideration** to avoid the robustness problem. One recommendation for such a situation is to keep the stability indices close to both ends $(\gamma_{n-1}, \gamma_2, \gamma_1)$ in the standard form, and choose the middle stability indices only larger than 1.5 times of the corresponding stability limits. Also, the stability indices are recommended to be between 1.5 and 4.

References

1. Lipatov AV, Sokolov NI (1978) Some sufficient conditions for stability and instability of continuous linear stationary systems. Translated from Automatika i Telemekhanika 9:30–37 (1978); Autom Remote Control (1979) 39:1285–1291
2. Manabe S (1999) Sufficient condition for stability and instability by Lipatov and its application to the coefficient diagram method. In: Proceedings of the 9th workshop on astrodynamics and flight mechanics, Sagamihara, ISAS, 22–23, pp 440–449
3. Franklin CF, Powell JD, Emami-Naeini A (2015) Feedback control of dynamic systems, 7th edn. Pearson Education Ltd., Essex, England
4. Franklin CF, Powell JD, Emami-Naeini A (1994) Feedback control of dynamic systems, 3rd edn. Addison-Wesley, Boston
5. Naslin P (1969) Essentials of optimal control. Illife books Ltd, London
6. Kessler C (1960) Ein Beitrag zur Theorie mehrschleifiger Regelungen. Regelungstechnik 8(8):261–266
7. Bhattacharyya SP, Chapellat H, Keel LH (1995) Robust control: the parametric approach. Prentice-Hall PTR, Upper Saddle River
8. Jayasuriya S, Song JW (1996) On the synthesis of compensators for nonovershooting step response. J Dyn Syst Meas Control-Trans ASME 118:757–763
9. Weinberg L (1963) Network analysis and synthesis. Macgraw-Hill, New York
10. Graham D, Lathrop RC (1953) The synthesis of optimum transient response: criteria and standard forms. AIEE Trans: pt II 72:273–288
11. Kitamori T (1979) A method of control system design based upon partial knowledge about controlled process. Trans SICE 15(4):549–555
12. Kim YC, Kim K, Manaba S (2006) Sensitivity of time response to characteristic ratios. IEICE Trans Fundam Electron, Commun Comput Sci E89-A(2):520–527

Chapter 3
Controller Design

Abstract In this chapter, the practical controller design procedures will be presented on the basis of the previous two chapters, "Introduction" and "Basics of Coefficient Diagram Method". Because practical design examples are also shown, ordinary design problems can be solved within this chapter. In **Interpretation of Design Specification**, the design specifications used in CDM are compared with the design specifications used in ordinary controller design. In **Definition of Basic Control Structure**, the methods for finding a controller structure that satisfies the design specifications will be explained. In **Interpretation of Robustness**, the interpretation of robustness in CDM will be presented, and it is compared with the usual interpretation of robustness. In **Design Process**, a brief explanation of the CDM design process will be given, where "Simultaneous Design of Controller and Characteristic Polynomial" is the central issue. As design examples, **Position Control**, **Position Control with Integrator**, and **PID Control** will be presented. In **Summary**, the important points are summarized.

3.1 Interpretation of Design Specification

The first step of the control system design is to specify the design specification. The specifications usually used in control system design are sometimes contradictory or insufficient. In many cases, they are neither sufficient nor necessary. **Specification of CDM** is given at first as a loose specification. Then, it is developed to a more concrete and consistent one after a trial-and-error approach during the design process. Thus, CDM specifications are difficult to make comparison with those specifications used in ordinary control design such as **Pole Placement**, **Time Response**, **Sensitivity/Complementary Sensitivity Functions**, and **Open-loop Transfer Function**. However, CDM design results approximately fulfill design specifications recommended in ordinary control design.

First, **Specification of CDM** is considered. Control system design is an effort to find a controller that satisfies stability, response, and robustness for the given plant. Stability is determined by the stability index of the characteristic polynomial. The choice of stability indexes close to the CDM standard form automatically guarantees

S. Manabe and Y. C. Kim, *Coefficient Diagram Method for Control System Design*, Intelligent Systems, Control and Automation: Science and Engineering 99,
https://doi.org/10.1007/978-981-16-0546-8_3

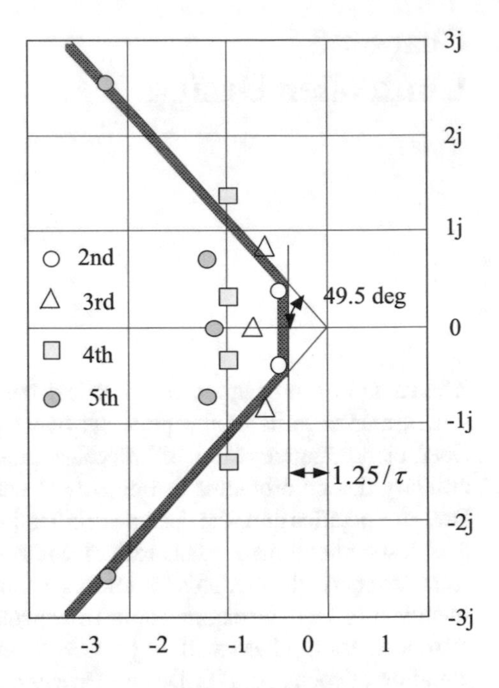

Fig. 3.1 Pole locations of standard CDM forms

stability. The response speed is uniquely defined by the equivalent time constant. Robustness is usually guaranteed by making the controller parameter non-negative. Thus, the specification of stability and robustness are automatically defined. The specification in which the designer has the choice is only the equivalent time constant. However, the plant usually has a natural response speed. When the specified equivalent time constant is consistent with such natural speed, the controller becomes a simple low-order one. If not, it becomes a high-order one with some complexity. Thus, the complexity of the controller and the equivalent time constant are in a trade-off relation.

The specification of CDM is summarized to the following three points. First, the stability indexes are to be chosen to the values close to the CDM standard form. Second, the controller parameters are chosen to be non-negative. Third, the equivalent time constant is to be chosen in consideration that it is in trade-off relation with controller complexity. The reason that the CDM specification takes such form is due to the salient feature of CDM design of "Simultaneous Design of Controller and Characteristic Polynomial".

In the specification of **Pole Placement**, the region of pole location is usually specified. In CDM standard form, the allowable region in s-plane is anywhere to the left of solid line in Fig. 3.1. The region is given by a vertical line $1.25/\tau$, where τ is equivalent time constant, and oblique lines of 49.5°. It is well known that the system with such pole location shows a good response.

As the specification of **Time Response**, the step response waveform is generally used. As shown in Fig. 3.2, the specifications are usually for rise time t_r, settling time

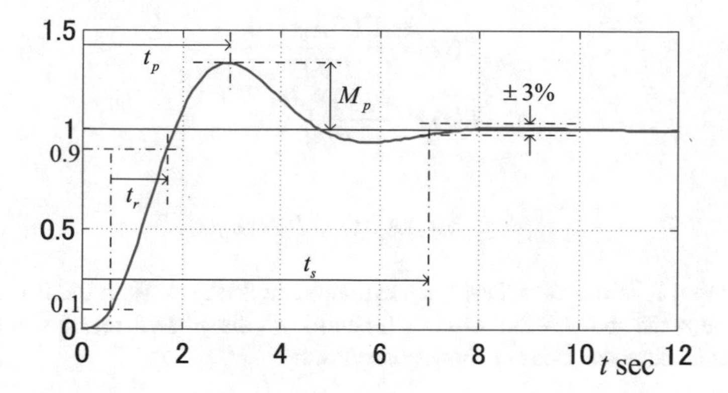

Fig. 3.2 Specification of step response

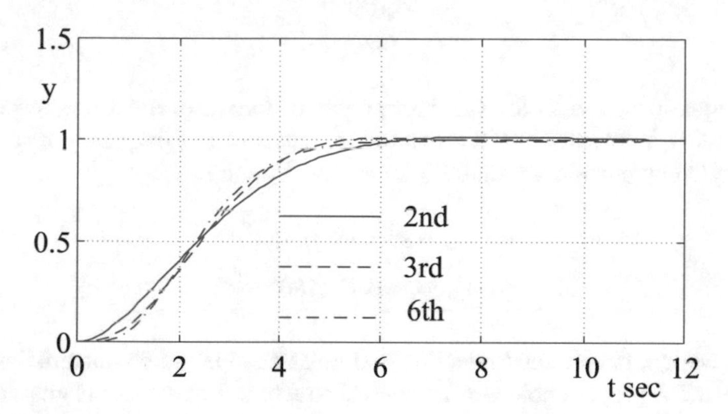

Fig. 3.3 Step response for standard CDM

t_s, overshoot M_p, and peak time t_p. For the settling time, the limit of convergence of 1% or 5% may be used instead of 3%. In practice, it is not necessary to make such a strict definition, because the settling time is considered as the time when the transient has roughly decayed. In CDM standard, the overshoot is 0% as shown in Fig. 3.3. The settling time is about 2.5–3 times the equivalent time constant. The CDM specification reflected on the time response is only equivalent time constant, which specifies the settling time. Other specifications of time response have to be accepted as they are as given by the design results.

The specifications of **Sensitivity/Complementary Sensitivity Functions** concern with robustness, disturbance rejection, and noise attenuation. The disturbance rejection characteristics are related to the low-frequency characteristics of sensitivity function, because the disturbance is of low-frequency nature. The noise attenuation characteristics are related to the high-frequency characteristics of complementary sensitivity function, because the noise is of high-frequency nature. Sensitivity function $S(s)$ and complementary sensitivity function $T(s)$ are defined as follows:

$$S(s) = \frac{A_c(s)A_p(s)}{P(s)}, \tag{3.1}$$

$$T(s) = \frac{B_c(s)B_p(s)}{P(s)}, \tag{3.2}$$

where

$$P(s) = A_c(s)A_p(s) + B_c(s)B_p(s). \tag{3.3}$$

Input–output relations for CDM were explained in Sect. 1.4. With the use of these relations and the above definitions of $S(s)$ and $T(s)$, disturbance rejection and noise attenuation characteristics are shown as follows:

$$y = \frac{B_p(s)}{A_p(s)} S(s)d, \tag{3.4}$$

$$y = T(s)(-n). \tag{3.5}$$

In these equations, y, d, and n stand for output, disturbance, and noise, respectively. Robustness is represented by the maximum values of the frequency responses of sensitivity/complementary sensitivity functions, Γ_s and Γ_c.

$$\Gamma_s = \max_{\omega} |S(j\omega)|, \tag{3.6}$$

$$\Gamma_c = \max_{\omega} |T(j\omega)|. \tag{3.7}$$

In CDM design, disturbance rejection and noise attenuation characteristics can be adjusted by the proper choice of the control structure. Robustness is guaranteed by the selection of controller parameters to be non-negative. In ordinary design, Γ_s and Γ_c become less than 1.5. These values roughly correspond to the results of the usual robust control design.

The specifications of **Open-loop Transfer Function** $G(s)$ are used in the classical control. They are disturbance rejection/noise attenuation characteristics, cross-over frequency, and phase/gain margins. Because $S(s) \simeq 1/G(s)$ in low-frequency range, and $T(s) \simeq G(s)$ in high-frequency range, the specifications of $G(s)$ for the low-frequency and high-frequency ranges can be obtained. At cross-over frequency range, the phase/gain margins give the specification of stability. At CDM design, the phase margin is around $\phi_m = 42°$, and gain margin is around $g_m = 3$.

When the maximum values of frequency responses of sensitivity/complementary sensitivity functions, Γ_s and Γ_c, are given, the frequency response of open-loop transfer function $G(s)$ must lie in the region shown in Fig. 3.4. Then, the phase/gain margins are given as follows:

$$\phi_m = 2\sin^{-1}(0.5/\Gamma_{min}), \tag{3.8}$$

$$g_{m1} \le G_{gm} \le g_m, \tag{3.9}$$

where $\Gamma_{min} = \min(\Gamma_s, \Gamma_c)$, $g_m = \Gamma_s/(\Gamma_s - 1)$, and $g_{m1} = (\Gamma_c - 1)/\Gamma_c$.

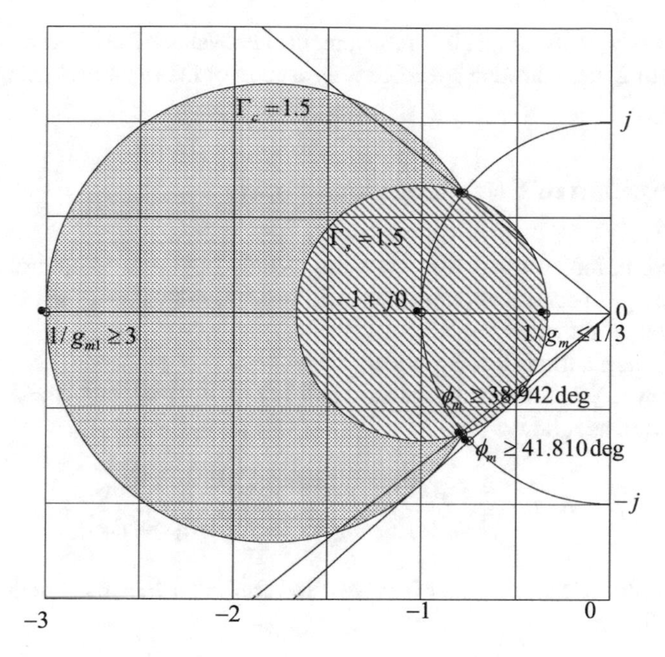

Fig. 3.4 Phase/gain margin based on Γ_s and Γ_c

In the above equation, G_{gm} is allowable variation of $|G(j\omega)|$ for $arg\ Gj\omega) = -180°$. Generally, the system becomes unstable for an increase in gain. However, for the system called a conditionally stable system, the system becomes unstable for a decrease in gain. For this reason, the lowest limit of gain variation is also given. The recommended values are as follows:

$$\Gamma_s = \Gamma_c = \Gamma_{min} = 1.5, \tag{3.10}$$

$$\phi_m = 2\sin^{-1}(0.5/\Gamma_{min}) = 38.942°, \tag{3.11}$$

$$1/3 \leq G_{gm} \leq 3. \tag{3.12}$$

As shown in Fig. 3.4, the phase margin is virtually $\phi_m = 41.810° \simeq 42°$. Because $|S(j\omega)| \leq \Gamma_s, |1 + G(j\omega)| \geq 1/\Gamma_s$. Thus, the locus of $G(j\omega)$ is outside of the circle with the center at -1 and the radius of $1/\Gamma_s$. Because $|T(j\omega)| = |G(j\omega)|/|1 + G(j\omega)| \leq \Gamma_c$, the locus of $G(j\omega)$ lies outside of the circle (the circle of Apollonius) whose diameter spans the two points, $-\Gamma_c/(\Gamma_c + 1)$ and $-\Gamma_c/(\Gamma_c - 1)$. The center of the circle is at $-\Gamma_c^2/(\Gamma_c^2 - 1)$, and the radius is $\Gamma_c/(\Gamma_c^2 - 1)$. From these relations, the phase/gain margins are obtained.

In short, the CDM design is an effort to find a proper control structure as well as a proper stability index and equivalent time constant. Because the specifications usually used in control system design retain only weak relations with the specifications used in CDM, it is very difficult to deduce the CDM specification from the ordinary control design specifications. It is more practical to evaluate the CDM design results by interpreting them through specifications used in ordinary control design.

3.2 Definition of Basic Control Structure

In this section, the basic control structure, namely the order of controller and the allowable range of controller parameters, will be briefly explained. The block diagram of the system is shown in Fig. 3.5.

At first, the relations among plants, controllers, and characteristic polynomials will be shown. When the order of the plant is m_p/n_p, the plant transfer function $G_p(s)$ is given as follows:

$$G_p(s) = \frac{B_p(s)}{A_p(s)} = \frac{n_{m_p}s^{m_p} + \cdots + n_1 s + n_0}{d_{n_p}s^{n_p} + \cdots + d_1 s + d_0}. \tag{3.13}$$

When the order of the controller is m_c/n_c, the controller transfer function $G_c(s)$ is given as follows:

$$G_c(s) = \frac{B_c(s)}{A_c(s)} = \frac{k_{m_c}s^{m_c} + \cdots + k_1 s + k_0}{l_{n_c}s^{n_c} + \cdots + l_1 s + l_0}. \tag{3.14}$$

The characteristic polynomial $P(s)$ is given as follows:

$$P(s) = A_c(s)A_p(s) + B_c(s)B_p(s) = a_n s^n + \cdots + a_1 s + a_0, \tag{3.15}$$

where

$$n = n_c + n_p, \quad a_n = l_{n_c} d_{n_p},$$

$$a_1 = l_1 d_0 + l_0 d_1 + k_1 n_0 + k_0 n_1, \quad a_0 = l_0 d_0 + k_0 n_0.$$

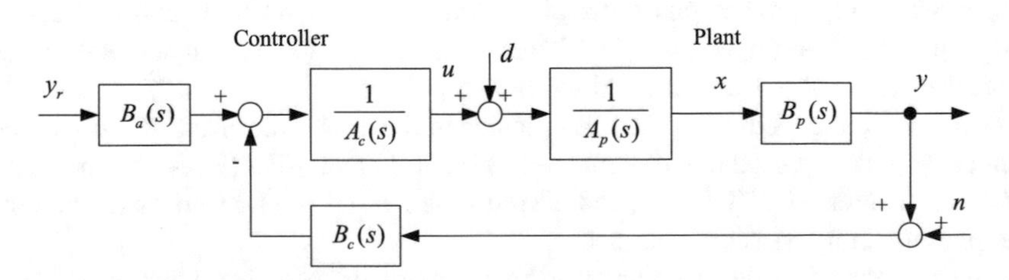

Fig. 3.5 Block diagram of control system in CDM

At the definition of control structure, crucial considerations have to be paid to the disturbance rejection and noise attenuation characteristics. They will be addressed in the following. The disturbance rejection characteristics $T_{yd}(s)$ are given as follows:

$$y = T_{yd}(s)d, \tag{3.16}$$

$$T_{yd}(s) = G_p(s)S(s) = \frac{B_p(s)A_c(s)}{P(s)}. \tag{3.17}$$

The steady state value of $T_{yd}(s)$ denoted as $T_{yd}(0)$ is given as follows:

$$T_{yd}(0) = \frac{n_0 l_0}{l_0 d_0 + k_0 n_0}. \tag{3.18}$$

From this result, it is clear that the condition $l_0 = 0$ is necessary to remove constant disturbance. The noise attenuation characteristics are related to the high-frequency characteristics of complementary sensitivity function $T(\infty)$.

$$y = T(s)(-n) = \frac{B_c(s)B_p(s)}{P(s)}(-n), \tag{3.19}$$

$$T(\infty) = \lim_{s\to\infty} \frac{k_{m_c} n_{m_p}}{l_{n_c} d_{n_p} s^{(n_c+n_p)-(m_c+m_p)}}. \tag{3.20}$$

The noise attenuation characteristics can be adjusted by the adjustment of the order of the denominator polynomial of the controller, n_c.

There are various kinds in the types of plants. Some are easy to control, although the order of the plant is high. Some are difficult to control even though the order is low. The low-order controllers are sufficient for the easy-to-control plants, but high-order controllers are necessary for difficult-to-control plant. It is recommended to start the design on the basis of the $1/1$ order, $2/2$ order, and $3/3$ controllers shown below.

$$G_c(s) = \frac{k_1 s + k_0}{s}. \tag{3.21}$$

$$G_c(s) = \frac{k_2 s^2 + k_1 s + k_0}{l_2 s^2 + s}. \tag{3.22}$$

$$G_c(s) = \frac{k_3 s^3 + k_2 s^2 + k_1 s + k_0}{l_3 s^3 + l_2 s^2 + s}. \tag{3.23}$$

In each controller, the lowest order parameter of the denominator, l_0, is made 0 in order to cancel the steady disturbance. In order to clarify the meaning of the controller parameters, normalization is made such that $l_1 = 1$. The $1/1$ order controller is the PI controller. The $2/2$ order controller looks to be the same as the PID controller, but it is different in that the parameter l_2 in the denominator is an important parameter with a strong influence on the overall control characteristics, and thus the number of controller parameters is four. For this reason, it is called generalized PID. In ordinary

PID, the number of parameters is three, because l_2 can be any value sufficiently small and is not expected to be the design parameter. In almost all plants, the 2/2 order controller (generalized PID) suffices. Only when it becomes truly necessary, the 3/3 order controller is used. When it is sought to augment the noise attenuation characteristics, it suffices to increase the order of the denominator polynomial of the controller, n_c, while the numerator is intact. The stability indexes can be kept at proper values when the denominator parameters are properly chosen. Because the controller parameters are basically non-negative, their lower bounds are 0. Their upper bounds are to be determined in consideration of the physical limitations of the controllers, and it is difficult to give general guidelines.

At the design, it is recommended to start with the low-order controller and move to the higher order controller step by step, when satisfactory results cannot be obtained. Such cases are when proper stability indexes are not obtained, or controller parameters become negative. In actual design, the order of controllers is properly assessed with reference to the past design experiences. After the order of the controller is roughly defined, it is further modified to satisfy the noise attenuation characteristics. In such a trial-and-error approach, the control structure is finally defined.

3.3 Interpretation of Robustness

In this section, the meaning of robustness will be clarified, and the specific considerations to attain robustness will be presented. First, it will be shown that **Robustness in CDM** is different from the ordinary definition and is more restrictive. Then, **Nature of Plants** will be discussed to show that there are cases where robustness cannot be attained even with the utmost effort in the controller design. Finally, it will be shown that there are cases where robustness is lost by **Improper Design**. The necessary precautions to avoid such situations will be presented.

As shown in Sect. 2.2 Coefficient Diagram, robustness in CDM is defined as "the robustness of the stability to the variation of the specific plant/controller parameters." However, poor robustness can occur in both cases, where the system is poor in stability or sufficient in it. The ordinary definition of robustness does not distinguish the two cases. In **Robustness in CDM**, a clear distinction is made for these two cases. It is restricted only to the case where stability is sufficient. In CDM, stability is uniquely determined by the stability index of the characteristic polynomial, which is the denominator term of sensitivity/complementary sensitivity function. However, robustness is determined by the numerator terms of the sensitivity/complementary sensitivity functions, which reflect the component polynomials of the characteristic polynomial. As shown in Sect. 2.2, robustness is deteriorated by the positive feedback of the controller, and robustness is improved by the negative feedback. Thus, the most effective means to achieve robustness is to keep the controller parameter non-negative. If positive feedback exists, there must be strong negative feedback to cancel this positive feedback. As a result, it is clearly noticed that the coefficients of the component polynomials are far above the coefficients of the characteristic polynomial in the coefficient diagram.

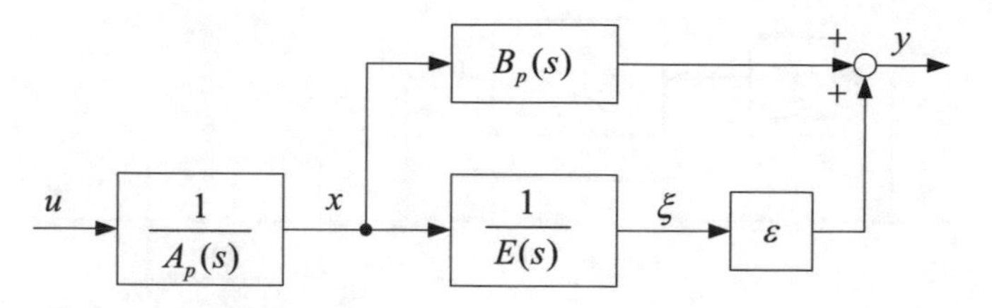

Fig. 3.6 Pole-zero cancellation in RPF

Next, the relation between robustness and **Nature of Plants** will be discussed. By experience, it is known that some plants are easy to control, while others are difficult. When the plant is easy to control, a sufficiently stable and robust system can be realized with a simple low-order controller. But when the plant is difficult to control, it is impossible to design a controller which satisfies stability and robustness simultaneously, even though the controller order is allowed to be very high. They become a trade-off relation. The effort to satisfy stability necessitates to conspicuously deteriorate robustness, and the effort to satisfy robustness necessitates to conspicuously deteriorate stability. In classical control, this problem is called pole-zero cancellation problem, and in modern control, it corresponds to the controllability/observability problem. First, the plant as shown in Fig. 3.6 is considered.

The plant is expressed in Right Polynomial Fraction(RPF) form as follows:

$$A_p(s)x = u, \qquad E(s)\xi = x, \qquad y = B_p(s)x + \epsilon\xi. \tag{3.24}$$

The transfer function expression for above RPF is as follows:

$$y = G_p(s)u, \tag{3.25}$$

$$G_p(s) = \left[B_p(s) + \tfrac{\epsilon}{E(s)}\right]\tfrac{1}{A_p(s)} = \tfrac{B_p(s)E(s)+\epsilon}{A_p(s)E(s)}. \tag{3.26}$$

With controller transfer function $G_c(s)$, the open-transfer function $G(s)$ is as follows:

$$G(s) = G_c(s)G_p(s) = \frac{B_c(s)[B_p(s)E(s)+\epsilon]}{A_c(s)A_p(s)E(s)}. \tag{3.27}$$

The characteristic polynomial $P(s)$ becomes

$$P(s) = [A_c(s)A_p(s) + B_c(S)B_p(s)]E(s) + B_c(s)\epsilon. \tag{3.28}$$

When ϵ becomes 0, pole-zero cancellation of $E(s)$ takes place in the open-loop transfer function. The characteristic polynomial becomes as follows:

$$P(s) = [A_c(s)A_p(s) + B_c(S)B_p(s)]E(s). \tag{3.29}$$

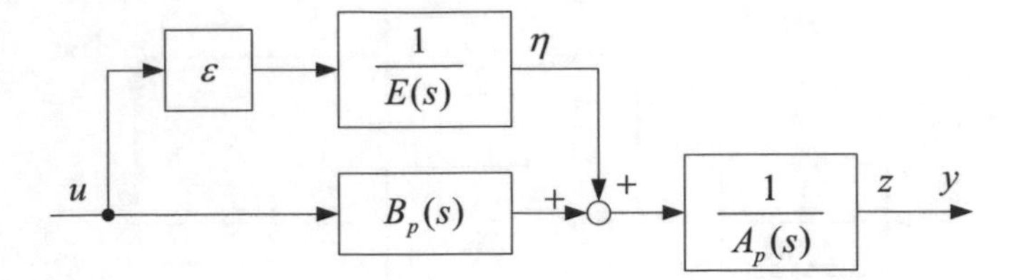

Fig. 3.7 Pole-zero cancellation in LPF

Thus, the character of $E(s)$ cannot be changed by the controller. When it contains oscillatory roots, small negative real roots, or unstable roots, a satisfactory closed-loop system cannot be realized. When ϵ is not 0, it is possible to realize the desired characteristic polynomial. However, in such cases, usually, the controller order becomes high and some controller parameters become negative. The robustness is conspicuously deteriorated. In the above discussion, the system with small negative real roots is considered as insufficient in stability in the sense that the convergence is too slow.

Next, let us consider the plant as shown in Fig. 3.7. The plant is expressed in Left Fraction Polynomial(LPF) form as follows:

$$A_p(s)z = B_p(s)u + \eta, \qquad E(s)\eta = \epsilon u, \qquad y = z. \tag{3.30}$$

The transfer function expression for above LPF is as follows:

$$y = G_p(s)u, \tag{3.31}$$

$$G_p(s) = \frac{1}{A_p(s)}\left[B_p(s) + \frac{\epsilon}{E(s)}\right] = \frac{B_p(s)E(s)+\epsilon}{A_p(s)E(s)}. \tag{3.32}$$

These are the same as Eqs. (3.25) and (3.26). Thus, the characteristic polynomial is the same as Eq. (3.28). The same discussion is possible.

At pole-zero cancellation, the plant becomes controllable–unobservable when it is expressed in RPF. In this expression, the input is connected to all state variables, but the output is not connected to all state variables. On the other hand, the plant becomes uncontrollable–observable when it is expressed in LPF. In this expression, the input is not connected to all state variables, but the output is connected to all state variables. In CDM, the pole-zero cancellation approach is preferred to controllability–observability commonly used in modern control. The first reason is that what actually exists is pole-zero cancellation condition, and controllability–observability depends on the interpretation of the state variables. The second reason is that, at pole-zero cancellation, such different cases as the pole and zero are exactly equal, slightly different, or far apart can be treated continuously with the same procedure. Because of these features, uncontrollable, difficult-to-control with poor robustness, or easy-to-control systems can be treated in the same manner.

Finally, the case where robustness is lost due to **Improper Design** will be discussed. Generally, each plant has a specific response speed. In ordinary control, negative feedback makes the closed-loop response faster than that speed. If it is desired to make it slow, some positive feedback is necessary. In such case, some controller parameter becomes negative, and robustness becomes poor. In such cases, it is required to specify the equivalent time constant smaller and to make the response faster. In CDM, the selection of equivalent time constant to the proper value is of utmost importance. However, it is very difficult to present the general theory, and it can be shown only by specific examples. With some experiences, the proper range of the equivalent time constant can be guessed by drawing some portion of the coefficient diagram. Then, by a CDM design program (CDM Toolbox in Appendix), the best equivalent time constant will be obtained by the systematic search.

The conclusions of the above analysis are as follows. First, the most effective way to attain good robustness is to keep the controller parameters non-negative. When negative parameters are required, careful bit-by-bit adjustment is requested. Second, for the system with pole-zero cancellation, it is impossible to design controllers, which satisfy robustness and stability condition simultaneously. In such cases, it is required to add new detectors, such that the plant is modified to favorable characteristics. Third, improper selection of equivalent time constant results in poor robustness. The most effective way to enhance the ability to select a proper equivalent time constant is to pile up the experiences of drawing coefficient diagrams.

3.4 Design Process

The outline of design process has been shown in Sect. 1.4 **Outline of Design Process**. In this section, a more detailed design process is presented. Because there are many types of plants, it is difficult to give a general design procedure. In order to give meaningful results in the design procedure, the plant is restricted to a stable plant with minimum-phase zero, which roughly corresponds to a plant with non-negative plant parameters. Also, such plants with some pole-zero cancellation, which have problems in terms of controllability/observability, are excluded. When the plant is restricted to this type, design can proceed by proper selection of stability indexes and equivalent time constant, where controller parameters are restricted to non-negative. The controller with non-negative parameter roughly corresponds to the stable and minimum-phase controller.

Generally, the stability indexes are recommended to be the CDM standard form. When the plant is difficult to control, where stability and robustness are in trade-off relation, the stability indexes are recommended to be chosen as in Sect. 2.6. They are as follows:

$$\gamma_{n-1} = \gamma_2 = 2, \quad \gamma_1 = 2.5, \quad \gamma_i > 1.5\gamma_i^*, \quad i = 3 \ldots n-2. \tag{3.33}$$

When the plants in consideration are fairly easy to control as assumed in this section, it is recommended that only the stability index of the highest order is chosen to be greater than 2, while the other stability indexes are chosen to be CDM standard form.

$$\gamma_{n-1} \geq 2, \quad \gamma_1 = 2.5, \quad \gamma_i = 2, \quad i = 2 \dots n-2. \tag{3.34}$$

Under this assumption, the general design procedure by CDM will be presented in the following. In order to help in understanding, a definite plant example, rather than general plants, is used. The example plant is the one used in Examples 7.21–7.24, 7.30, 7.31, Chap. 7 ofFranklin's textbook [1], or in Examples 7.29–7.31, Franklin [2]. By the use of the same example, the difference between CDM and other design approaches will be clarified.

First, **Relation between Orders of Plant and Controller** is discussed. Then, the **Plant** used in the examples is shown, and the design results in the reference are briefly explained. Then, the design is made for the simplest **1/1 order controller**. For such a simple design, the graphical design by the coefficient diagram is most effective. Finally, more complex **2/2 order controller** is designed. Such a controller can be given various capabilities, but the design process becomes more complex. The design is made with the help of CDM Toolbox.

The **Relation between Orders of Plant and Controller** is discussed in the following. The order of the plant is m_p/n_p and that of the controller is m_c/n_c. Then, the order of the characteristic polynomial is $n_p + n_c$. The characteristic polynomial has $n_p + n_c + 1$ coefficients. Because the characteristic polynomial retains the same characteristics even when divided by a constant, its freedom is $n_p + n_c$. The controller has $n_c + m_c + 2$ parameters. Because one parameter can be normalized to 1 without changing its characteristics, the freedom is $n_c + m_c + 1$. When the freedom of the characteristic polynomial coincides with that of the controller, any kind of characteristic polynomial can be designed by the proper choice of controller parameters. This condition becomes $m_c = n_p - 1$, or the order of the numerator of the controller is less than that of the denominator of the plant by 1.

$$n_p + n_c = n_c + m_c + 1, \quad m_c = n_p - 1. \tag{3.35}$$

In actual design, due to limitations in the selection of controller parameters, the design freedom is lost accordingly. The n freedom of the nth-order characteristic polynomial is represented by n parameters consisting of $n-1$ stability indexes, γ_i, and the equivalent time constant, τ. The two important limitations in the design are the low-frequency response of the open-loop transfer function (low-frequency response of sensitivity function) and the robustness requirement that all controller parameters are non-negative. Because design freedom is lost due to such limitations, the condition is relaxed in such a way that the low-order stability indexes are chosen as the CDM standard form, but the high-order ones are relaxed as greater than 2, and the equivalent time constant is chosen accordingly. In ordinary cases, the proper design is made by the proper selection of the highest order stability index and the equivalent time constant. In a similar manner, the design of a low-order controller, such that $m_c < m_p - 1$, can be effectively made. Such flexibility is the unique feature of CDM.

Next, the **Plant** used in the examples will be given. The transfer function of the plant is a 0/3 order as follows:

$$G_p(s) = \frac{B_p(s)}{A_p(s)} = \frac{10}{s(s+2)(s+8)} = \frac{10}{s^3+10s^2+16s}. \tag{3.36}$$

For this plant, a 2/3 order and 2/2 order controllers are designed. In these examples, some controllers produce systems with a lack of robustness due to improper choice of the equivalent time constant (Examples 7.21, 7.22, 7.30, 7.31). In *Inward Approach* [3] used in modern control, the proper selection of the equivalent time constant is of utmost importance, but it is often overlooked and becomes the serious pitfall of the design.

Because the plant is 0/3 order, the 2/2 order controller has sufficient design freedom. In actual CDM design, however, it is recommended to start from a low-order controller, because some plant is easy to control and a low-order controller may produce satisfactory results. Thus, **1/1 order controller** is first considered. This is as Eq. (3.21) in Sect. 3.2.

$$G_c(s) = \frac{k_1 s + k_0}{s}. \tag{3.37}$$

The characteristic polynomial becomes fourth order as follows:

$$\begin{aligned} P(s) &= s(s^3+10s^2+16s) + 10(k_1 s + k_0) \\ &= s^4 + 10s^3 + 16s^2 + 10k_1 s + 10k_0 \\ &= a_4 s^4 + a_3 s^3 + a_2 s^2 + a_1 s + a_0. \end{aligned} \tag{3.38}$$

Because the freedom of the controller is 2, and that of the characteristic polynomial is 4, the design freedom is short by 2. Thus, design is made by specifying $\gamma_2 = 2$ and $\gamma_1 = 2.5$, and accepting the design results as they are for γ_3 and τ. The design results are as follows:

$$\gamma_3 = \frac{a_3^2}{a_4 a_2} = \frac{10^2}{1 * 16} = 6.25. \tag{3.39}$$

$$\gamma_2 = \frac{a_2^2}{a_3 a_1} = \frac{16^2}{10 * (10k_1)} = 2. \tag{3.40}$$

$$\Rightarrow \quad k_1 = 1.28, \quad a_1 = 12.86.$$

$$\gamma_1 = \frac{a_1^2}{a_2 a_0} = \frac{12.8^2}{16 * (10k_0)} = 2.5. \tag{3.41}$$

$$\Rightarrow \quad k_0 = 0.4096, \quad a_0 = 4.096.$$

$$\tau = a_1/a_0 = 12.8/4.096 = 3.125. \tag{3.42}$$

These results are expressed in a coefficient diagram as in Fig. 3.8. In the lower part, the CDM-type Bode diagram is shown. The detailed procedure of drawing the diagram is explained in Sect. 3.5. The detailed design results are obtained by CDM Toolbox with these controller parameters, as shown in Fig. 3.9. The summary of the design results is as follows:

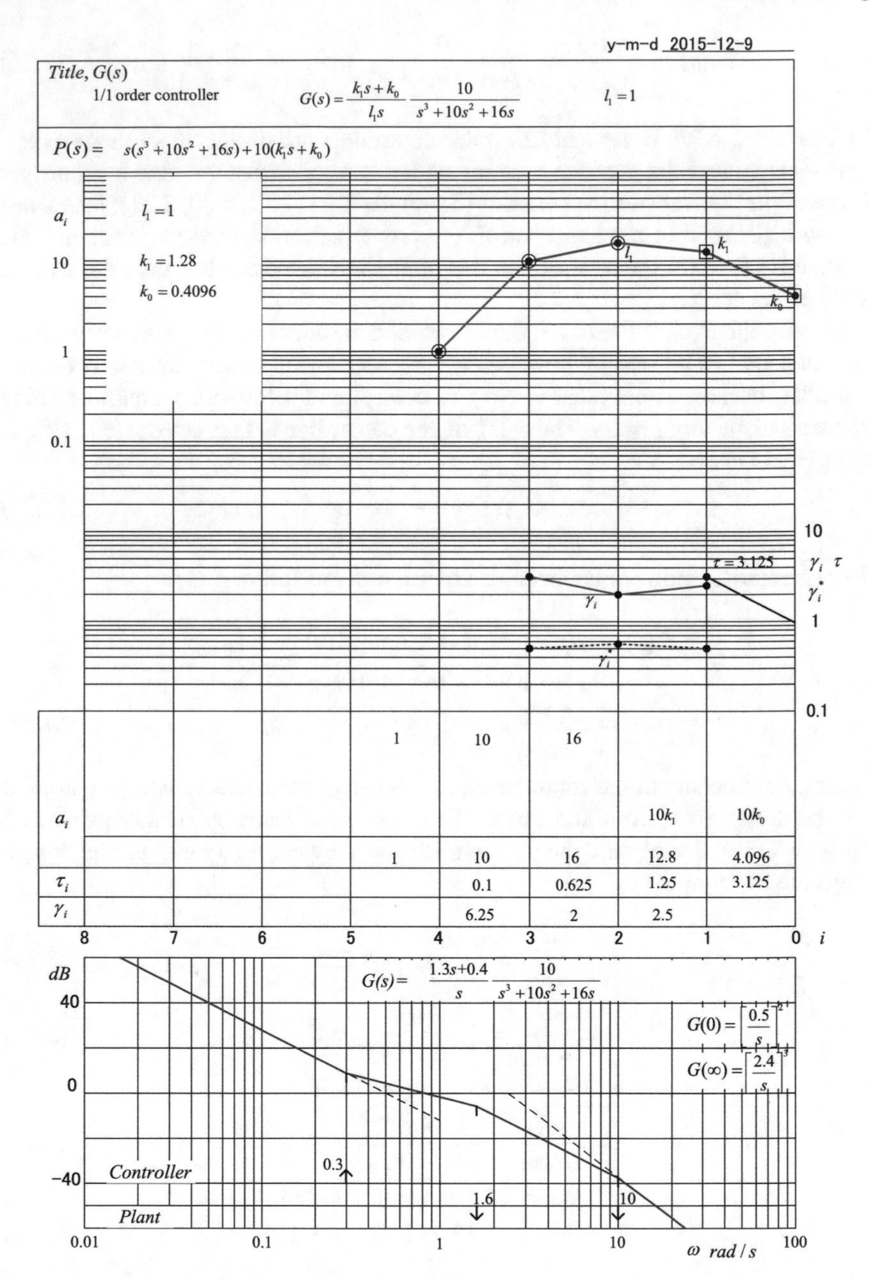

Fig. 3.8 Coefficient diagram, 1/1 order controller

$$G(s) = G_c(s)G_p(s) = \frac{1.28s + 0.4096}{s} \frac{10}{s^3 + 10s^2 + 16s}, \tag{3.43}$$

$$\begin{aligned} P(s) &= s^4 + 10s^3 + 16s^2 + 12.8s + 4.096, \\ \gamma_i &= [6.25 \;\; 2 \;\; 2.5], \quad \tau = 3.125, \\ s_i &= -8.2393, \quad -0.5579 \pm j0.67975, \quad -0.6411, \\ \phi_m &= 40.708^\circ, \quad g_m = 10. \end{aligned}$$

In this design, a hand-written design with CDM design form is very effective, where the calculations of Eq. (3.39)–(3.42) are made systematically. From the coefficients, $a_4 = 1$, $a_3 = 10$, $a_2 = 16$, the equivalent time constants of high order, $\tau_3 = a_4/a_3 = 0.1$, $\tau_2 = a_3/a_2 = 0.625$, are obtained. Then $\gamma_3 = \tau_2/\tau_3 = 6.25$ is obtained. From $\gamma_2 = 2, \tau_1 = \gamma_2\tau_2 = 1.25$ is obtained. From $\gamma_1 = 2.5, \tau = \gamma_1\tau_1 = 3.125$ is obtained. With these results, the low-order coefficients of the characteristic polynomial are obtained, such as $a_1 = a_2/\tau_1 = 12.8$, and $a_0 = a_1/\tau = 4.096$. Then, controller parameters are obtained such as $k_1 = a_1/10 = 1.28$ and $k_0 = a_0/10 = 0.4096$.

In the calculation to obtain Fig. 3.9 by CDM Toolbox, the following command is used.

```
>> ap=[1 10 16 0];bp=10;ac=[1 0];bc=[1.28 0.4096];tm=0.5; cc
```

In these commands, symbols ap, bp, ac, bc correspond to $A_p(s)$, $B_p(s)$, $A_c(s)$, and $B_c(s)$. The symbol tm specifies time scale of time response, usually chosen to be 0.5 or 1. When it is smaller, time scale is expanded. The command cc computes the various characteristics of the system for the given controller and plant. The symbols ba, aa, g, tau, gs, rr correspond to $B_a(s)$, $P(s)$, τ, γ_i, γ_i^*, s_i. The s_i are closed-loop poles. The symbol pmgm shows the phase margin ϕ_m and the gain margin g_m in the combined form. The symbol wpmgm specifies the frequencies in rad/s for such margins. In the response figures, $G(s)$, $S(s)$, *and* $T(s)$ mean the open-loop transfer function, the sensitivity function, and the complementary sensitivity function. The $W(s)$ is the transfer function for command following.

$$S(s) = \frac{A_p(s)A_c(s)}{P(s)}, \quad T(s) = \frac{B_p(s)B_c(s)}{P(s)}, \quad W(s) = \frac{B_p(s)B_a(s)}{P(s)}. \tag{3.44}$$

In this design, the design becomes complete in the process of building the coefficient diagram by simply filling the CDM design form as in Fig. 3.10 with hand. In the coefficient diagram, the curvature is the measure of stability. Thus, the ratio of length and width has to be fixed. In the CDM design form, 1 decade in length and the order scale in width are set to 2 cm. It is recommended to magnify the form Fig. 3.10 to this scale.

Because $\tau = 3.125$ in this case, and τ is within 1–10, there is no problem for the ordinate being a_i. However, when τ is very small, it is very awkward to use a_i. The ordinate is modified as $a_i(10)^i$, $a_i(100)^i$, etc. Similarly, when τ is very large, the

```
>> ap=[1 10 16 0]; bp=[10]; ac=[1 0]; bc=[1.28 0.4096]; tm=0.5; cc

ba =
    0.4096
bc =
   1.28    0.4096
ac =
   1     0
aa =
   1.0    10.0    16.0    12.8    4.096
g =
   6.25     2.0     2.5
tau =
     3.125
gs =
   0.50     0.56     0.50
rr =
  -8.2393
  -0.5598 + 0.6798i
  -0.5598 - 0.6798i
  -0.6411
pmgm =
      40.7079   10.00
wpmgm =
        0.7969    3.5777
```

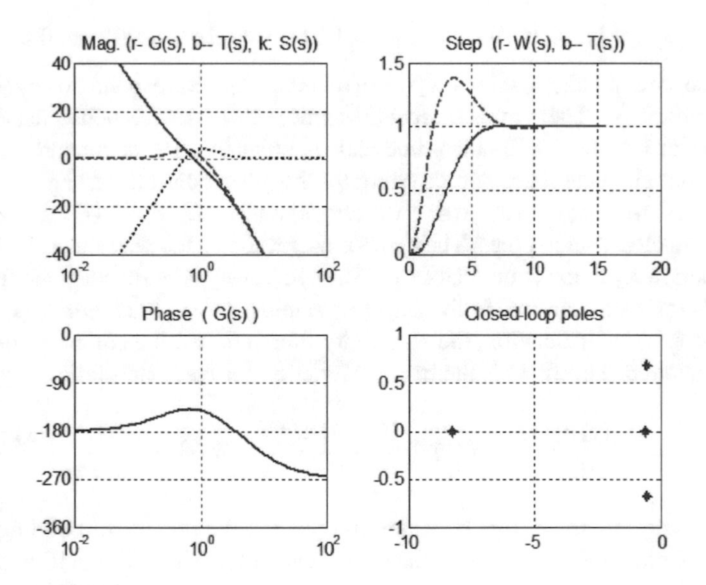

Fig. 3.9 Design results, 1/1 order controller

ordinate is modified as $a_i(0.1)^i$, $a_i(0.01)^i$, etc. With such modification, the shape of the coefficient diagram becomes round and more understandable.

Next, the **2/2 order controller** is considered. For the 1/1 order controller, the equivalent time constant is 3.125 s. But for the 2/2 order controller, a shorter equivalent time constant with a faster response is expected. The controller is as follows:

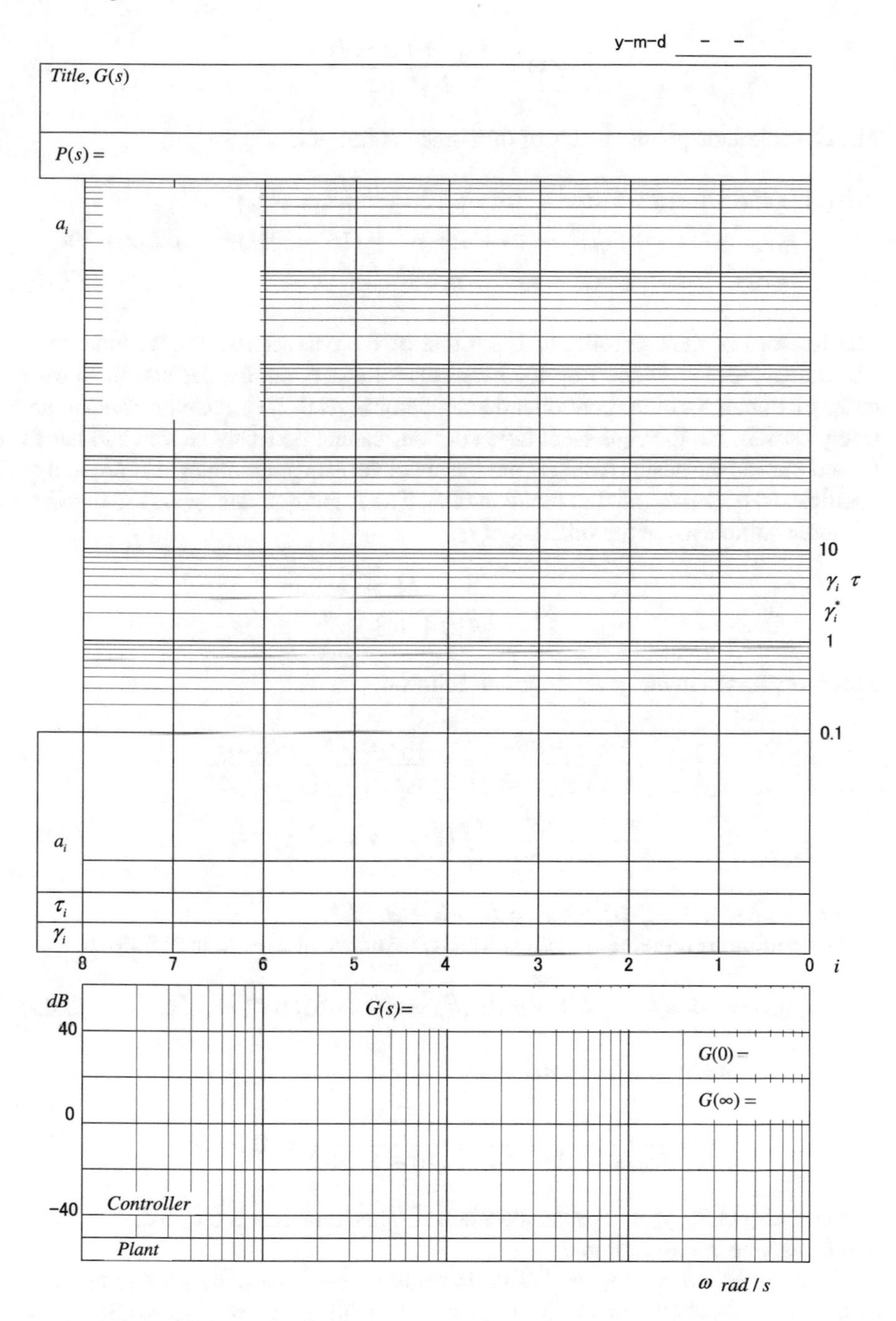

Fig. 3.10 CDM design form

$$G_c(s) = \frac{k_2 s^2 + k_1 s + k_0}{l_2 s^2 + s}. \tag{3.45}$$

The characteristic polynomial is of fifth order as follows:

$$\begin{aligned} P(s) &= (l_2 s^2 + s)(s^3 + 10s^2 + 16s) + 10(k_2 s^2 + k_1 s + k_0) \\ &= l_2 s^5 + (1 + 10l_2)s^4 + (10 + 16l_2)s^3 + (16 + 10k_2)s^2 + 10k_1 s + 10k_0 \\ &= a_5 s^5 + a_4 s^4 + a_3 s^3 + a_2 s^2 + a_1 s + a_0. \end{aligned} \tag{3.46}$$

The freedom of the controller is 4, but that of the characteristic polynomial is 5. The design freedom is short by 1. This result follows naturally, because the lowest order parameter, l_0, of the controller denominator is set 0. Then, stability indexes are freely chosen, but the equivalent time constant cannot be freely chosen and has to be accepted as the design result. From the characteristic polynomial, it is found that coefficients, a_5, a_4, a_3, are the function of l_2. The stability index, γ_4, is the function of these coefficients, or the function of l_2.

$$\gamma_4 = \frac{(1 + 10l_2)^2}{l_2(10 + 16l_2)}. \tag{3.47}$$

This is expressed in the general form as follows:

$$\begin{aligned} \gamma_4 &= \frac{(d_3 l_1 + d_2 l_2)^2}{d_3 l_2 (d_2 l_1 + d_1 l_2)} = \frac{d_3^2 l_2^2 (l_1/l_2 + d_2/d_3)^2}{d_3 d_2 l_2^2 (l_1/l_2 + d_1/d_2)} \\ &= \frac{d_3 (x + A)^2}{d_2 x} = \frac{d_3}{d_2}(x + 2A + A^2/x), \end{aligned} \tag{3.48}$$

where $x = (l_1/l_2 + d_1/d_2)$ and $A = (d_2/d_3 - d_1/d_2)$.

The minimum value, γ_{4min}, of γ_4, which is attained at $x = A$, is as follows:

$$\gamma_{4min} = 4d_3 A/d_2 = 4(1 - d_3 d_1/d_2^2) = 4(1 - 16/100) = 3.36. \tag{3.49}$$

From $x = A$ and $l_1 = 1$, l_2 is obtained.

$$l_2 = \frac{l_1}{(d_2/d_3 - 2d_1/d_2)} = \frac{1}{(10 - 32/10)} = 0.14706. \tag{3.50}$$

Because $\gamma_4 \geq 3.36$, $\gamma_4 = 2$ cannot be attained. This limitation is due to the condition that l_2 must be the real number.

From the above results, the design is made systematically by filling the CDM design form, as shown in Fig. 3.11. First, the stability indexes, $\gamma_4 = 3.36,\ \gamma_3 = \gamma_2 = 2,\ \gamma_1 = 2.5$, are filled. Then, the high-order coefficients of the characteristic polynomial, $a_5 = l_2 = 0.14706,\ a_4 = 1 + 10l_2 = 2.4706,\ a_3 = 10 + 16l_2 = 12.353$, are filled. Then, the equivalent time constants of various orders are obtained

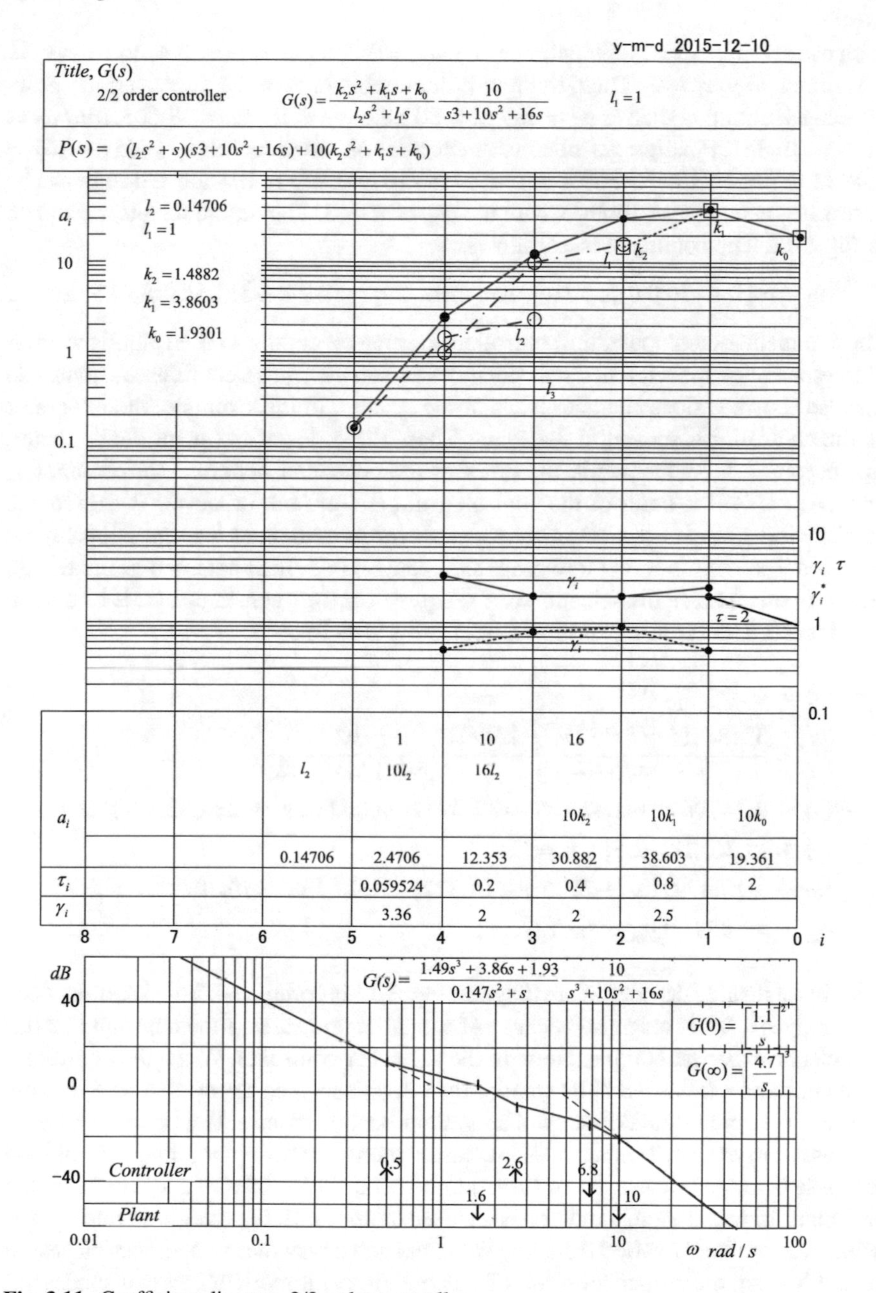

Fig. 3.11 Coefficient diagram, 2/2 order controller

such as $\tau_4 = a_5/a_4 = 0.059524$, $\tau_3 = a_4/a_3 = 0.2$, $\tau_2 = \gamma_3\tau_3 = 0.4$, $\tau_1 = \gamma_2\tau_2 = 0.8$, *and* $\tau = \gamma_1\tau_1 = 2$. Then, the low-order coefficients of the characteristic polynomial are obtained such as $a_2 = a_3/\tau_2 = 30.882$, $a_1 = a_2/\tau_1 = 38.603$, *and* $a_0 = a_1/\tau = 19.361$. Finally, controller parameters are obtained such as $k_2 = (a_2 - 16) = 1.4882$, $k_1 = a_1/10 = 3.8603$, *and* $k_0 = a_0/10 = 1.9361$. The above design can be performed using CDM Toolbox with the use of $\tau = 2$. The design results are shown in Fig. 3.12. The command is as follows:

```
>> ap=[1 10 16 0 0];bp=10; nc=1;mc=2;gr=[2 2 2 2.5];t=2;tm=0.5; gc
```

The command gc computes the controller parameters for the specified plant, the order of the controller, the reference stability indexes, and the equivalent time constant. The gc also shows various characteristics of the system. In this example, the integrator of the controller is moved to the plant. Then, the order of the plant denominator, ap, becomes 4. As the result, the order of the controller denominator becomes 1, or nc=1, while the order of the controller numerator is 2, or mc=2. For reference stability indexes, gr, $\gamma_4 = 2$ is given. However in command gc, the Diophantine equation is solved from the low-order side, and $\gamma_4 = 3.36$ is obtained as the design result. In this design, the selection of the equivalent time constant, t=τ, is of crucial importance. The summary of the design results is as follows.

$$
\begin{aligned}
G(s) &= G_c(s)G_p(s) \\
&= \frac{1.4882s^2 + 3.8603s + 1.9301}{0.14706s^2 + s}\frac{10}{s^3 + 10s^2 + 16s}, \\
P(s) &= 0.14706s^5 + 2.4706s^4 + 12.353s^3 + 30.882s^2 + 38.603s + 19.301, \\
\gamma_i &= [3.36\ \ 2\ \ 2\ \ 2.5], \quad \tau = 2, \\
s_i &= -10.479, \quad -1.7236 \pm j1.6322, \quad -1.4368 \pm j0.39793, \\
\phi_m &= 42.989^\circ, \quad g_m = 6.6727.
\end{aligned}
\tag{3.51}
$$

With these results, the coefficient diagram Fig. 3.11 is completed. This 2/2 order controller gives a faster response because of a smaller equivalent time constant and has sufficient robustness comparable with the 1/1 order controller. When the equivalent time constant τ is chosen to be smaller than 2, γ_4 becomes larger than 3.36, but the robustness becomes deteriorated. The system becomes unstable at $\tau = 1$. When τ is chosen larger than 2, robustness increases to unnecessarily larger values, and the response becomes slower. From these points, The choice of $\tau = 2$ seems to be the optimum. In this design, another trial-and-error approach is possible to find $\tau = 2$, where τ is gradually varied from any value in such a way that γ_4 becomes closest to 2. In any case, the proper selection of τ is the utmost important key to a successful CDM design. However, it is very difficult to show the general approach applicable to various circumstances, and example designs will be shown instead.

```
>> ap=[1 10 16 0 0];bp=[10];nc=1;mc=2;gr=[2 2 2 2.5];t=2;tm=0.5; gc

ba =
    1.9301
bc =
    1.4882    3.8603    1.9301
ac =
    0.1471    1.0000
aa =
    0.1471    2.4706    12.3529    30.8824    38.6029    19.3015
g =
    3.3600    2.0000    2.0000    2.5000
tau =
    2
gs =
    0.5000    0.7976    0.9000    0.5000
rr =
  -10.4792 + 0.0000i
   -1.7236 + 1.6322i
   -1.7236 - 1.6322i
   -1.4368 + 0.3979i
   -1.4368 - 0.3979i
pmgm =
   42.9886    6.6727
wpmgm =
    1.8262    6.7473
```

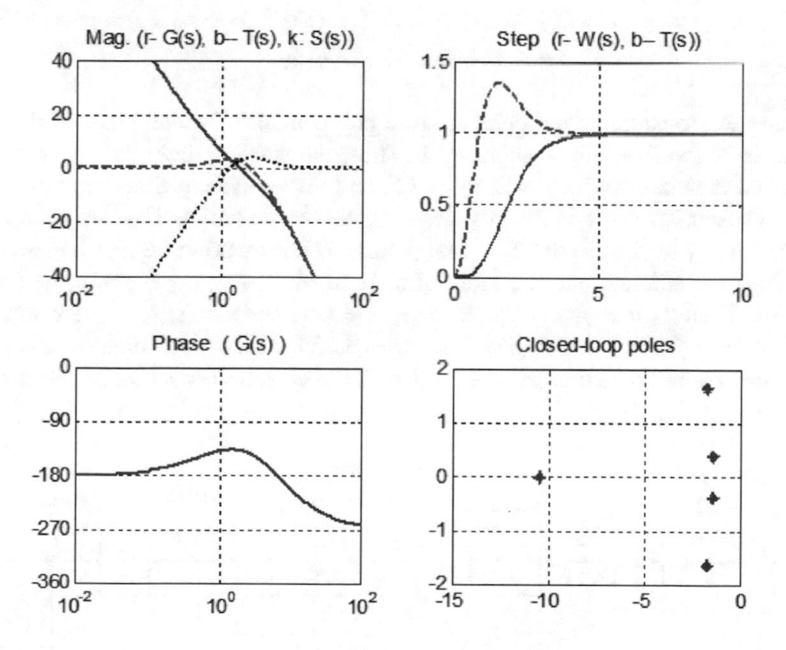

Fig. 3.12 Design results, 2/2 order controller

3.5 Position Control

As a simple example of controller design, the position control with velocity feedback is considered. The control system is shown in Fig. 3.13.

The denominator/numerator polynomials for $T_v = 0.25$, $T_m = 1$ are given as follows:

$$A_p(s) = (T_v s + 1)(T_m s + 1)s = 0.25s^3 + 1.25s^2 + s, \quad B_p(s) = 1. \tag{3.52}$$

The controller is of 1/0 order, where it is assumed that the velocity is measured.

$$A_c(s) = l_0 = 1, \quad B_c(s) = k_1 s + k_0, \quad B_a(s) = k_0. \tag{3.53}$$

The open-loop transfer function is as follows:

$$G(s) = G_c(s)G_p(s) = \frac{k_1 s + k_0}{1} \frac{1}{0.25s^3 + 1.25s^2 + s}. \tag{3.54}$$

The characteristic polynomial is as follows:

$$\begin{aligned} P(s) &= 0.25s^3 + 1.25s^2 + (1 + k_1)s + k_0 \\ &= a_3 s^3 + a_2 s^2 + a_1 s + a_0. \end{aligned} \tag{3.55}$$

Because the freedom of the characteristic polynomial is 3, while that of the controller is 2, the freedom is short by 1. Thus, the stability indexes are chosen as the CDM standard form,$\gamma_2 = 2$, $\gamma_1 = 2.5$, and the equivalent time constant is not freely chosen. Its value is directly given as the design result. The design is made systematically by the use of CDM design form as illustrated in Fig. 3.14. First, $G(s)$ and $P(s)$ are entered, and the table of a_i is filled. Then $\gamma_2 = 2\, and\, \gamma_1 = 2.5$ are entered. Then $\tau_2 = a_3/a_2 = 0.2$, $\tau_1 = \tau_2\gamma_2 = 0.4, and\, \tau = \tau_1\gamma_1 = 1$ are entered. Then $a_1 = a_2/\tau_1 = 3.125\, and\, a_0 = a_1/\tau = 3.125$ are entered. Finally, controller parameters are obtained as $k_1 = a_1 - 1 = 2.125\, and\, k_0 = a_0 = 3.125$. With these

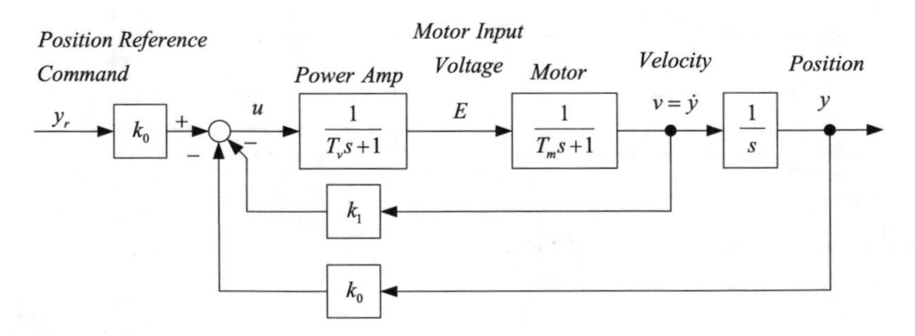

Fig. 3.13 Position control

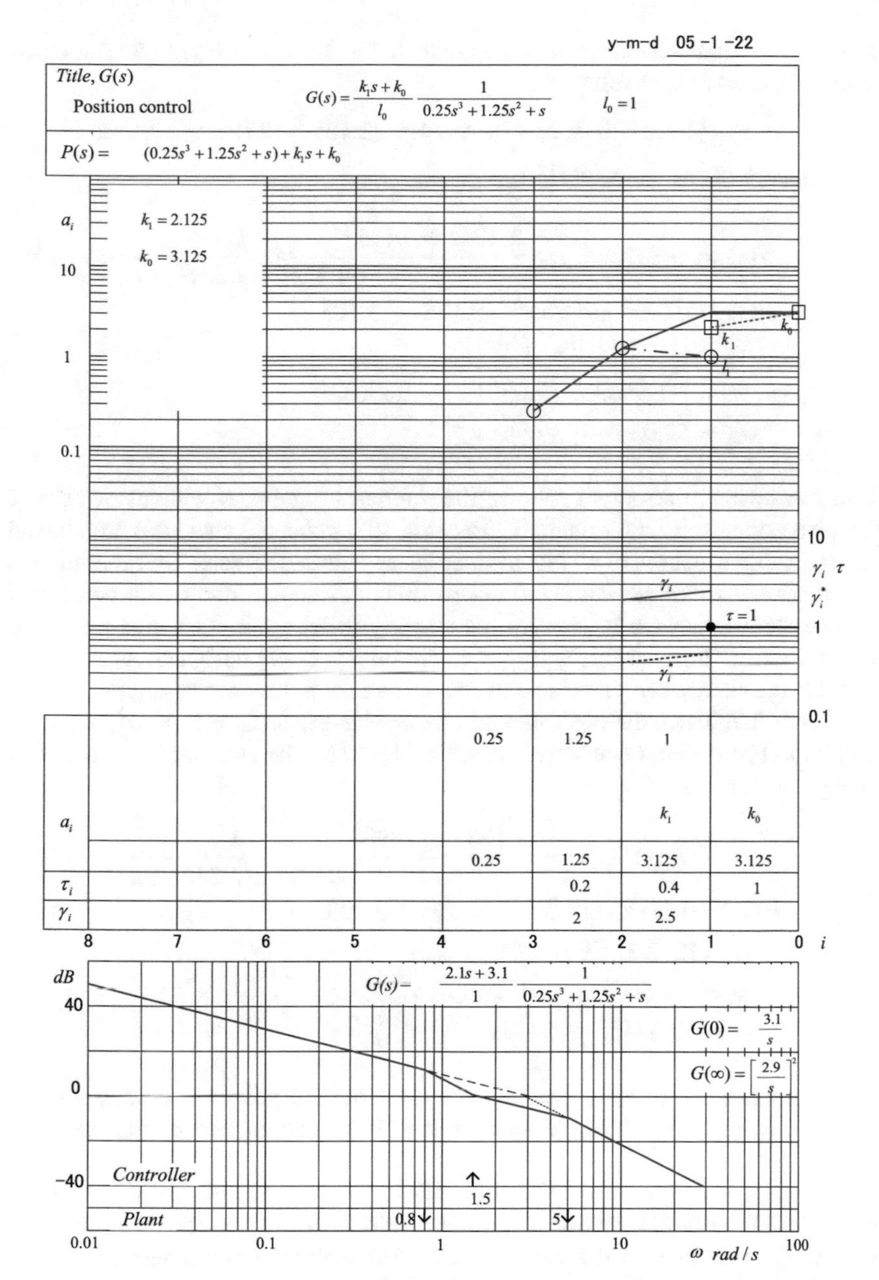

Fig. 3.14 Coefficient diagram, position control

results, the various characteristics are computed as shown in Fig. 3.15, The CDM Toolbox command is as follows:

```
>> ap=[1 1.25 16 0];bp=1;ac=1;bc=[2.125 3.125];tm=0.5; cc
```

The summary of design results is as follows:

$$\begin{aligned} G(s) &= G_c(s)G_p(s) = \frac{2.125s + 3.125}{1} \frac{1}{0.25s^3 + 1.25s^2 + s}, \\ P(s) &= 0.25s^3 + 1.25s^2 + 3.125s + 3.125, \\ \gamma_i &= [2 \ \ 2.5], \quad \tau = 1, \\ s_i &= -1.5568 \pm j2.0501, \quad -1.8863, \\ \phi_m &= 52.935^\circ, \quad g_m = \infty. \end{aligned} \tag{3.56}$$

In this design, the output follows the reference command without overshoot. The phase margin is larger than 50°. Because all 3 poles (eigenvalues) are aligned vertically, the responses of all eigenvalues decay at the same rate. This condition is sometimes called as *eigenvalue coalescence*. In this example, although the equivalent time constant is not freely chosen, the stability indexes are chosen as the CDM standard form. Thus, a stable and robust controller is easily designed. As shown in Sect. 2.5, the case, when there is no overshoot and the poles exactly align vertically, is for $\gamma_1 = 2.7$. Then, the controller parameter changes as $k_0 = a_0 = a_1^2/(a_2\gamma_1) = 2.893519$. The design results are shown in Fig. 3.16. The summary of the design results is as follows:

$$\begin{aligned} G(s) &= G_c(s)G_p(s) = \frac{2.125s + 2.8935}{1} \frac{1}{0.25s^3 + 1.25s^2 + s}, \\ P(s) &= 0.25s^3 + 1.25s^2 + 3.125s + 2.8935, \\ \gamma_i &= [2 \ \ 2.7], \quad \tau = 1.08, \\ s_i &= -1.6667 \pm j2.0412, \quad -1.6667, \\ \phi_m &= 55.344^\circ, \quad g_m = \infty. \end{aligned} \tag{3.57}$$

As the design results show, there is no overshoot, and poles are exactly aligned vertically. Even in the third-order system, $\gamma_1 = 2.5$ is chosen, because it is easier to remember.

Finally, the way to draw the CDM-type Bode diagram is explained. For the controller design, only the coefficient diagram suffices. However, the Bode diagram is effective to understand the design results. Although the Bode diagram is shown in CDM Toolbox, it is the results only. The CDM-type Bode diagram is important because it clarifies the effect of each parameter on the overall characteristics. The drawing process is different from the ordinary way in order to keep close relations with the coefficient diagram. As an example, Fig. 3.13 is used.

First, the open-loop transfer function $G(s)$ is entered in the CDM design form. Because it is used only for drawing, two-digit numbers suffice.

```
>> ap=[0.25 1.25 1 0]; bp=[1]; ac=[1]; bc=[2.125 3.125]; tm=0.5;  cc

ba =
   3.1250
bc =
   2.1250    3.1250
ac =
    1
aa =
   0.2500    1.2500    3.1250    3.1250
g =
   2.0000    2.5000
tau =
    1
gs =
   0.4000    0.5000
rr =
  -1.5568 + 2.0501i
  -1.5568 - 2.0501i
  -1.8863 + 0.0000i
pmgm =
  1.0e+03 *
   0.0529    8.6074
wpmgm =
   2.0807  270.5883
```

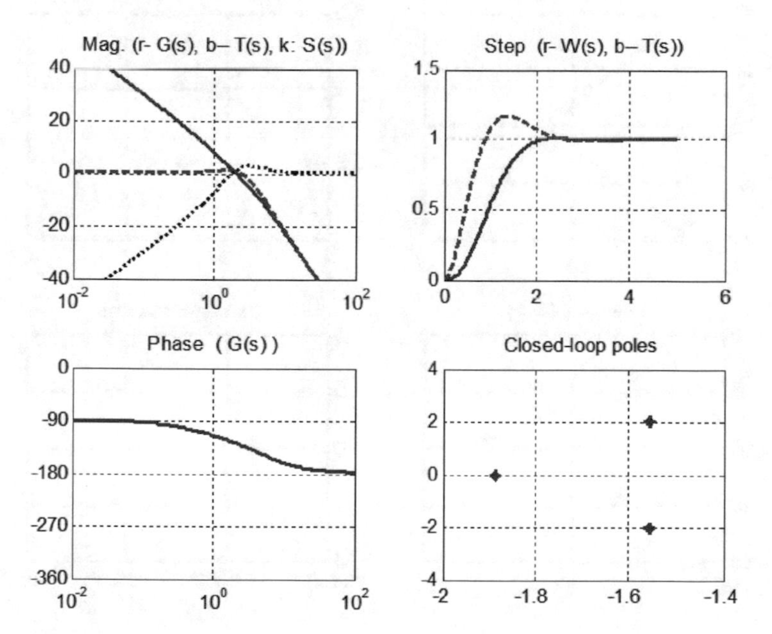

Fig. 3.15 Design results, position control

```
>> ap=[0.25 1.25 1 0]; bp=[1]; ac=[1]; bc=[2.125 2.893519]; tm=0.5;  cc

ba =
   2.8935
bc =
   2.1250   2.8935
ac =
    1
aa =
   0.2500   1.2500   3.1250   2.8935
g =
   2.0000   2.7000
tau =
   1.0800
gs =
   0.3704   0.5000
rr =
  -1.6667 + 2.0412i
  -1.6667 - 2.0412i
  -1.6667 + 0.0000i
pmgm =  1.0e+03 *
   0.0553   8.6069
wpmgm =
   2.0428  270.5820
```

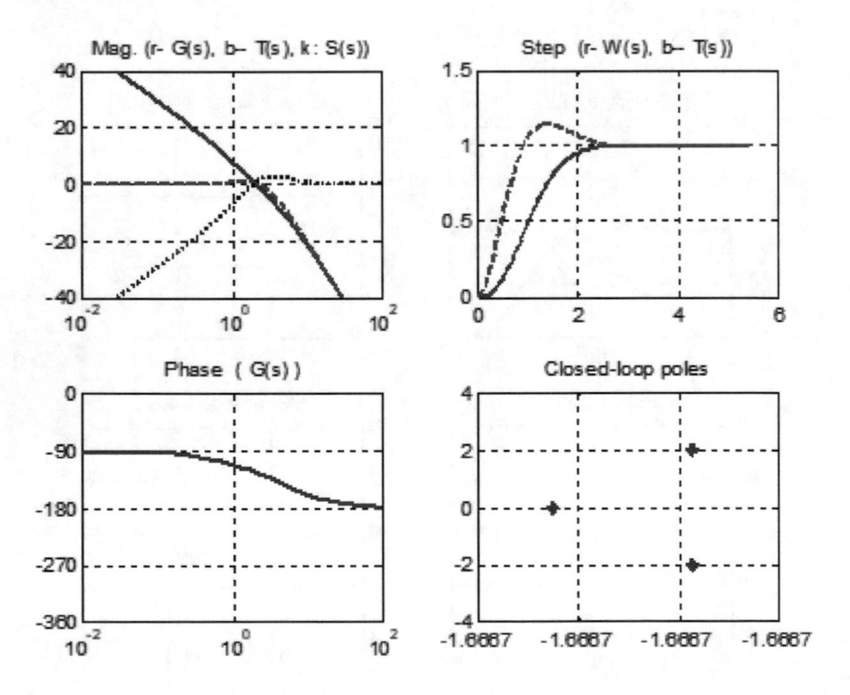

Fig. 3.16 Design results, position control with $\gamma_1 = 2.7$

$$G(s) = \frac{2.1s + 3.1}{1} \frac{1}{0.25s^3 + 1.25s^2 + s}. \tag{3.58}$$

Then, $G(0) = 3.1/s$, where $G(0)$ is $G(s)$ for sufficiently small s, is entered. Also, $G(\infty) = 2.1/(0.25s^2) = (2.9/s)^2$, where $G(\infty)$ is $G(s)$ for sufficiently large s, is entered. In the CDM-type Bode diagram, the break points are not poles or zeros but the ratios of the adjacent two parameters. For oscillatory poles or zeros, the square roots of parameters separated by two order are used. For the controller, upward break point is at $3.1/2.1 = 1.5$ with no downward break point. For the plant, two downward break points are at $1/1.25 = 0.8$ and $1.25/0.25 = 5$ with no upward break point. Then, an upward break point is marked on the line of *Controller*, and two downward break points on the line of *Plant*. In order to draw the CDM-type Bode diagram, first draw the $G(0) = 3.1/s$ line, which is a $-20°/decade$ line passing the point of 3.1 rad/s, 0 dB. Then, at 0.8 rad/s, bend it downward to $-40°/decade$. Then, at 1.5 rad/s, bend it to upward to $-20°/decade$. Finally, at 5 rad/s, bend it to downward to $-40°/decade$. This line should exactly coincide with the line $G(\infty) = (2.9/s)^2$, or the $-40°/decade$ line passing the point of 2.9 rad/s, 0 dB.

3.6 Position Control with Integrator

In this section, the position control with integrator is considered, where the integrator for disturbance compensation is added to the position control with velocity feedback in the previous section. The design is made for a concrete physical model, and the various problems, which arise in the actual design, are also considered. First, a brief description of **Plant** is given. Then **Controller** is explained. In the preparation of the controller design, the plant parameters are calculated. Then, the design proceeds for the case of **Controller with Integrator**, where the integrator in the controller is in operation. Finally, the design is made for the case of **Controller with Integrator Saturation**, where the integrator is out of operation.

First, the **Plant** is explained. This control system is a hand-made toy position control system for the students to understand basic control system characteristics. In this system, a toy DC motor drives a paper disk through a gearbox. The angular position is measured by a potentiometer. The signal is fed to analog operational amplifiers and power transistors, which drive the motor in such a way that the angular position matches to the position reference command. Even by such a simple system, students can feel the effects of the changes of the velocity and position feedback parameters on the response speed or the stability. The students can physically feel the hunting oscillation, which actually occurs in any real control system.

The plant consists of the motor drive, power amplifier, velocity sensor, and preamplifier. The equation of the motor drive is as follows:

$$J s\Omega_L = GR\, k_T i + d_1, \tag{3.59}$$
$$V_T = (R_a + R_1)i + E, \tag{3.60}$$
$$s\theta_L = \Omega_L, \tag{3.61}$$
$$V_{R1} = R_1 i,\ E = k_V\, GR\, \Omega_L,\ J = J_M\, GR^2 + J_L,\ GR = N_2/N_1. \tag{3.62}$$

The $J_M,\ J_L,\ J$ are the motor/disk/total moment of inertias. The GR is the gear ratio. The $k_T,\ k_V$ are motor torque/voltage constants. The R_a is the internal resistance of the motor, where the internal inductance is neglected. The R_1 is the resistance connected to the motor, where the voltage drop between the terminals, V_{R1}, is used to detect the motor current i. The d_1 is the external load torque, where the positive direction is defined as the direction for the torque to increase the velocity. This convention is adopted to harmonize with that of CDM. The $\Omega_L,\ \theta_L$ are the angular velocity/position of the disk.

The angular position, θ_L, is directly measured by the potentiometer. However the angular velocity, Ω_L, is calculated from the motor counter Electro-Motive Fsorce (EMF), E, which is estimated from the motor terminal voltage, V_T, and the voltage drop, V_{R1}, as shown in the following equations.

$$\Omega_L = \frac{E}{k_V GR} = \frac{1}{k_V GR}\left[V_T - \frac{R_a + R_1}{R_1} V_{R1}\right]. \tag{3.63}$$

When J is replaced by the time constant, T_m, in Eq. (3.59), the following equation is obtained.

$$T_m s\Omega_L = \frac{R_a + R_1}{k_V GR} i + d, \tag{3.64}$$

where

$$T_m = \frac{R_a + R_1}{k_V k_T GR^2} J, \qquad d = \frac{R_a + R_1}{k_V k_T GR^2} d_1.$$

The motor current, i, is supplied by the output voltage, V_T, of the power amplifier. In ordinary conditions, the output voltage, V_T, follows the input voltage, V_{T0}. However, due to safety consideration, the output voltage, V_T, is made lower than the input voltage, V_{T0}, in order to keep the motor current within the current limit, i_{LIM}, when the absolute value of the current is to exceed the limit. The following equation is obtained, if the counter-EMF, E, is eliminated from Eq. (3.59), and the above relation is used.

$$\begin{aligned} \frac{R_a + R_1}{k_V GR} i &= \left(\tfrac{V_{T0}}{k_V GR} - \Omega_L\right)\left[IF\ |\tfrac{V_{T0}}{k_V GR} - \Omega_L| < \tfrac{R_a + R_1}{k_V GR} i_{LIM}\right] \\ &\pm \left(\tfrac{R_a + R_1}{k_V GR} i_{LIM}\right)\left[IF\ |\tfrac{V_{T0}}{k_V GR} - \Omega_L| \geq \tfrac{R_a + R_1}{k_V GR} i_{LIM}\right]. \end{aligned} \tag{3.65}$$

In the above equation, $[IF\ Statement]$ is defined as a number, which is 1 if the *Statement* is true, and 0 if it is false. The input voltage, V_{T0}, is the output voltage of the preamplifier. The preamplifier is represented by a time-lag circuit, whose time constant is T_v, and the input is u.

$$(T_v s + 1)\frac{V_{T0}}{k_V GR} = u. \tag{3.66}$$

From the above equations, the equation of the plant, consisting of the motor drive, power amplifiers, velocity sensor, and preamplifier, is obtained. When the current is not saturated, the equation is as follows:

$$(T_v s + 1)(T_m s + 1)\Omega_L = u + (T_v s + 1)d, \tag{3.67}$$

$$s\theta_L = \Omega_L. \tag{3.68}$$

When the current is saturated, the equations become as follows:

$$(T_v s + 1)\frac{V_{T0}}{k_V GR} = u, \tag{3.69}$$

$$(T_m s + 1)\Omega_L = \pm\frac{R_a + R_1}{k_V GR} i_{LIM} + d, \tag{3.70}$$

$$s\theta_L = \Omega_L. \tag{3.71}$$

In this case, the plant output, $y = \theta_L$, is not connected to the plant input, u.
Next, the **Controller** is explained. It is a PID controller as follows:

$$u = -k_2 s y + k_1(y_r - y) + u_1, \tag{3.72}$$

$$s u_1 = k_0(y_r - y), \tag{3.73}$$

$$u_1 = u_1[IF\ |u_1| < u_{1LIM}] \pm u_{1LIM}[IF\ |u_1| \geq u_{1LIM}], \tag{3.74}$$

$$y = \theta_L, \quad sy = \Omega_L. \tag{3.75}$$

As the design base of controller, the plant parameters are calculated as follows:

$$\begin{aligned} GR &= 64.8, \quad J_M = 2.53125 \times 10^{-7}\,\text{kg m}^2, \quad J_L \simeq 0, \\ J &= GR^2 J_M, \quad R_a + R_1 = 1.2\,\Omega, \\ k_T &= 0.00218\,\text{Nm/A}, \quad k_V = 0.00218\,V/(\text{rad/s}), \\ T_m &= \frac{1.2 \times 2.53125 \times 10^{-7}}{0.00218^2} = 0.063915 \simeq 0.064, \\ T_v &= 0.016. \end{aligned} \tag{3.76}$$

The control system is shown in Fig. 3.17. In this figure, the saturation is indicated by a circle-dot symbol on the line with a note for the saturation values. In the controller design, the linear system is first considered, where the nonlinear elements such as the current saturation and integrator saturation are not present. This case is the design of **Controller with Integrator**. The denominator/numerator polynomials of the plant

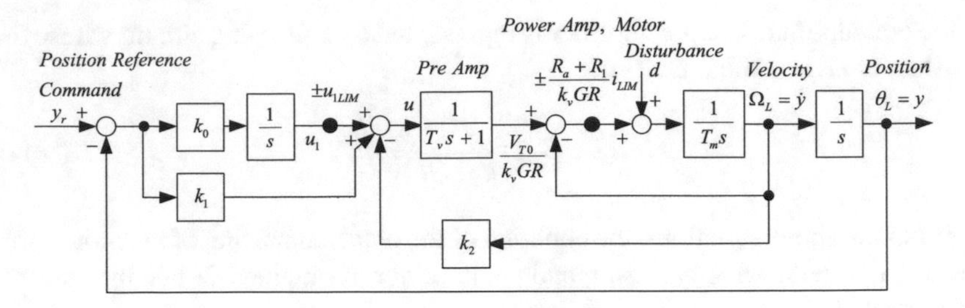

Fig. 3.17 Position control with integrator

are given as follows, where the parameters are known to be $T_v = 0.016 and\ T_m = 0.064$.

$$A_p(s) = (T_v s + 1)(T_m s + 1)s = 0.001024s^3 + 0.08s^2 + s, \tag{3.77}$$

$$B_p(s) = 1. \tag{3.78}$$

The controller is the 2/1 order controller as shown below where the condition that the velocity sensor is available as well as the position sensor is utilized.

$$A_c(s) = s, \quad B_c(s) = k_2 s^2 + k_1 s + k_0, \quad B_a(s) = k_1 s + k_0. \tag{3.79}$$

The open-loop transfer function is as follows:

$$G(s) = G_c(s)G_p(s) = \frac{k_2 s^2 + k_1 s + k_0}{s} \frac{1}{0.1024s^3 + 0.08s^2 + s}. \tag{3.80}$$

The characteristic polynomial is as follows:

$$\begin{aligned} P(s) &= 0.001024s^4 + 0.08s^3 + (1 + k_2)s^2 + k_1 s + k_0 \\ &= a_4 s^4 + a_3 s^3 + a_2 s^2 + a_1 s + a_0. \end{aligned} \tag{3.81}$$

Although the same design procedure as in the previous section can be used in this design, an easier design approach by the use of CDM Toolbox is shown here. First, the stability indexes are somewhat modified as $\gamma_3 = 2$, $\gamma_2 = \gamma_1 = 2.5$. This modification is made in order to remove the overshoot from the response with integrator saturation. The equivalent time constant of the third order is as follows:

$$\tau_3 = a_4/a_3 = 0.0128. \tag{3.82}$$

Then, the equivalent time constant is obtained as follows:

$$\tau = \gamma_3 \gamma_2 \gamma_1 \tau_3 = 0.16. \tag{3.83}$$

The gc command used in Sect. 3.4 cannot be used here, because $B_a(s) = k_1 s + k_0$ and not k_0. Instead, g2c and c2g commands are used here.

```
>> ap=[0.001024 0.08 1 0 0];bp=1;nc= 0;mc=2;gr=[2 2.5 2.5];t=0.16;
>> [bc,ac,aa,g,tau,gs,rr]=g2c(ap,bp,nc,mc,gr,t,0);
>> ba=[bc(2) bc(3)];tm=0.5;
>> ba,bc,ac,
>> [aa,g,tau,gs,rr,pmgm,wpmgm]=c2g(ap,bp,ac,bc,ba,tm)
```

The summary of design results is as follows:

$$\begin{aligned} G(s) &= G_c(s)G_p(s) \\ &= \frac{2.125s^2 + 48.828s + 305.18}{s} \frac{1}{0.001024s^3 + 0.08s^2 + s}, \\ P(s) &= 0.001024s^4 + 0.08s^3 + 3.125s^2 + 48.828s + 305.18, \\ \gamma_i &= [2 \ \ 2.5 \ \ 2.5], \quad \tau = 0.16, \\ s_i &= -27.146 \pm j28.824, \quad -11.916 \pm j6.9356, \\ \phi_m &= 49.942^\circ, \quad g_m = \infty. \end{aligned} \tag{3.84}$$

The coefficient diagram is shown in Fig. 3.18, where the ordinate is $a_i \times 10^i$ instead of a_i. With this modification, the diagram becomes easier to draw and understand. The design results are shown in Fig. 3.19.

Now, the case for **Controller with Integrator Saturation** will be considered. This case corresponds to the condition $k_0 = 0$, and the controller is the 1/0 order controller as shown below.

$$A_c(s) = 1, \quad B_c(s) = k_2 s + k_1, \quad B_a(s) = k_1, \tag{3.85}$$

$$k_2 = 2.125, \quad k_1 = 48.828. \tag{3.86}$$

The design is made by the following command.

```
>> ap=[0.001024 0.08 1 0];bp=1;ac= 1;bc=[2.125 48.828];tm=0.5; cc
```

The summary of the design results is as follows:

$$\begin{aligned} G(s) &= G_c(s)G_p(s) = \frac{2.125s + 48.828}{1} \frac{1}{0.001024s^3 + 0.08s^2 + s}, \\ P(s) &= 0.001024s^3 + 0.08s^2 + 3.125s + 48.828, \\ \gamma_i &= [2 \ \ 2.5], \quad \tau = 0.064, \\ s_i &= -24.325 \pm j32.033, \quad -29.474, \\ \phi_m &= 52.935^\circ, \quad g_m = \infty. \end{aligned} \tag{3.87}$$

The design results are shown in Fig. 3.20.

The step responses for these cases are shown in Fig. 3.21. The solid line indicates the case for $k_0 = 305.18$, and the dashed line for $k_0 = 0$. The solid line is the same

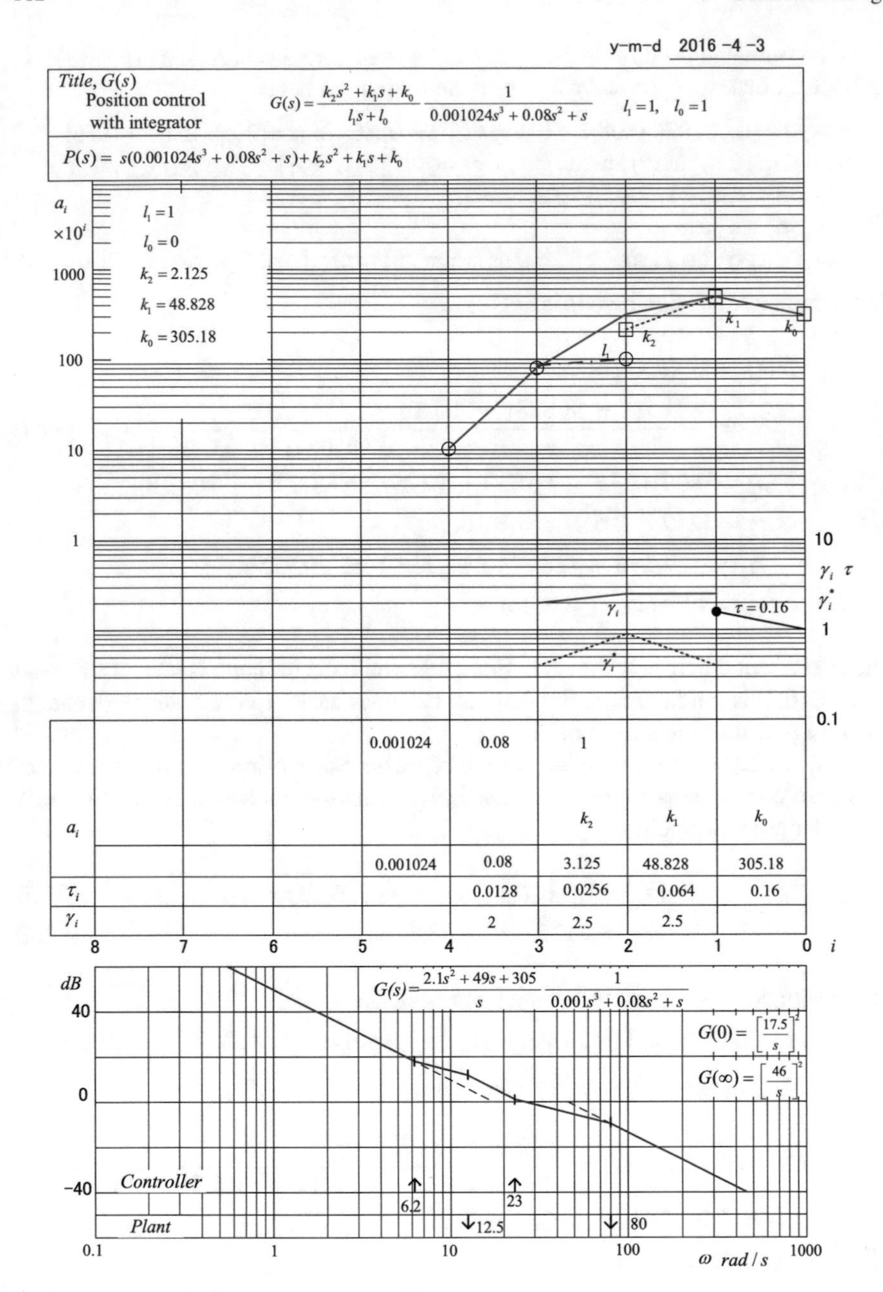

Fig. 3.18 Coefficient diagram, position control with integrator

```
>>ap=[0.001024 0.08 1 0 0];bp=[1];nc=0; mc=2;gr=[2 2.5 2.5]; t=0.16;
>>[bc,ac,aa,g,tau,gs,rr]=g2c(ap,bp,nc,mc,gr,t,0);
>>ba=[bc(2) bc(3)]; tm=0.5; ba, bc, ac,
>>[aa,g,tau,gs,rr,pmgm,wpmgm]=c2g(ap,bp,ac,bc,ba,tm)

ba =
  48.8281  305.1758
bc =
   2.1250   48.8281  305.1758
ac =
    1
aa =
   0.0010    0.0800    3.1250   48.8281  305.1758
g =
   2.0000    2.5000    2.5000
tau =
   0.1600
gs =
   0.4000    0.9000    0.4000
rr =
 -27.1464 +28.8239i
 -27.1464 -28.8239i
 -11.9161 + 6.9356i
 -11.9161 - 6.9356i
pmgm =
  49.4929   52.2157
wpmgm =
  30.1538  363.9521
```

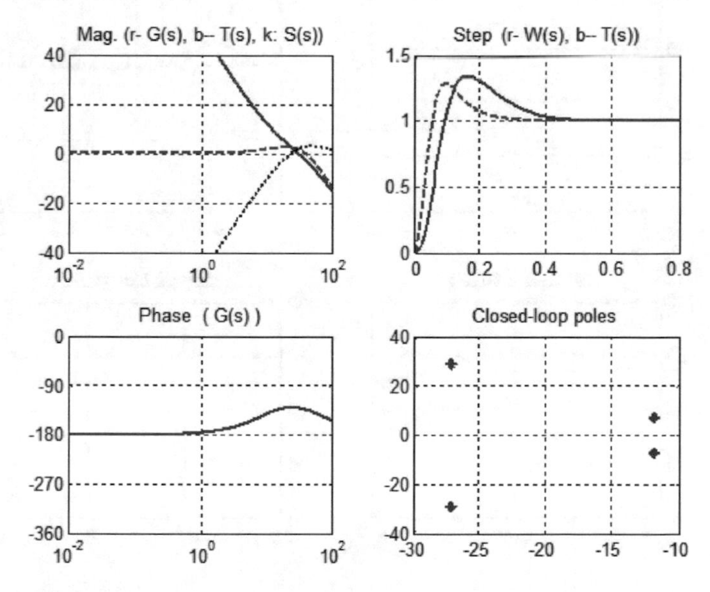

Fig. 3.19 Design results, position control with integrator

```
>> ap=[0.001024 0.08 1 0];bp=[1];ac=1; bc=[2.125 48.828]; tm=0.5; cc

ba =
   48.8280
bc =
    2.1250   48.8280
ac =
     1
aa =
    0.0010    0.0800    3.1250   48.8280
g =
    2.0000    2.5000
tau =
    0.0640
gs =
    0.4000    0.5000
rr =
 -24.3255 +32.0326i
 -24.3255 -32.0326i
 -29.4740 + 0.0000i
pmgm =
   52.9352   51.1596
wpmgm =
   32.5108  355.5003
```

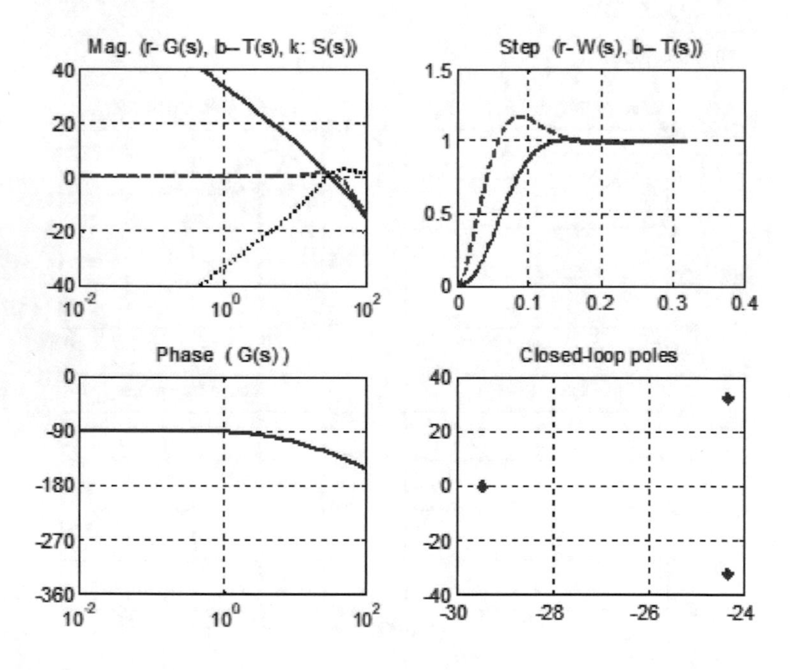

Fig. 3.20 Design results, position control with integrator saturation

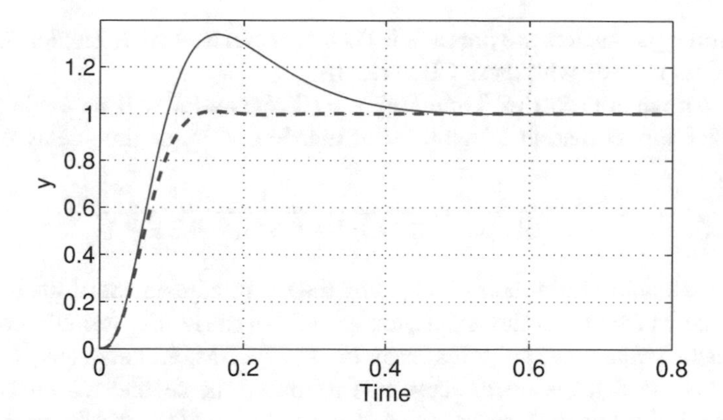

Fig. 3.21 Step responses: solid line with integrator, and dashed line with integrator saturation

as the response of $W(s)$ in Fig. 3.19. The dashed line is the response of $W(s)$ in Fig. 3.20. Although the actual response can be estimated from these figures, the nonlinear simulation is required to obtain an accurate response.

In this example, the saturation of the motor current is inevitable. Then, the system becomes open-loop temporarily, and the integrator continues to integrate. The integrator output becomes very large, and this phenomenon is the so-called Integrator Wind-up. To avoid this, the integrator output is to be limited to a reasonable saturation level. Because the integrator loses the control function under this condition, it is required for the reference command to be given directly, not through the integrator, to the controller. For this reason, the reference numerator polynomial is to be made as $B_a(s) = k_1 s + k_0$. When the integrator is saturated and the output is kept constant, the corresponding stability index is to be modified to avoid overshoot, such that $\gamma_2 = 2.5$. As explained above, some modification is necessary even to the linear part in actual design in order to cope with the nonlinearity inherent to the system. In this example, the modifications are $B_a(s) = k_1 s + k_0$ and $\gamma_2 = 2.5$.

3.7 PID Control

In this section, the design method of the PID controller is discussed. The conceived plant is a multistage time-lag plant. Such plant is approximated by a time delay with a time-lag. The time delay cannot be expressed in polynomial, and CDM design is not directly applicable. For this reason, **Approximation of Time Delay by Polynomial** is first discussed. Because the response of PID control is usually slow and analog control is difficult to implement, digital control is common and the system should be discretized. The CDM is applicable only to a continuous-time system. Thus, it is necessary to approximate the discrete-time system by the continuous system. **Approximation of Discrete-Time System** is next discussed. It is already well known that such approximation is possible by the addition of proper time delay to the system. Then **Design of PID Control** is made by CDM. In this process, the general formulas for controller parameters are presented. Then **Comparison with Ziegler & Nichols Approaches** is made with these CDM results.

First, **Approximation of Time Delay by Polynomial** will be explained. The polynomial approximation of the transfer function, e^{-Ls}, of time-delay of L is as follows:

$$e^{-Ls} \simeq \frac{1}{A_L(s)} = \frac{1}{0.1L^3s^3 + 0.5L^2s^2 + Ls + 1}. \tag{3.88}$$

This approximation formula is obtained by the minor adjustment of the third-order term of the Maclaurin series expansion of e^{Ls}. Because this formula consists of denominator polynomial only, it is easy to use in CDM, and also easy to remember. This is the first reason to adopt this formula. The second reason is that the approximation is accurate up to $\omega = 1/L$ rad/s. At $s = j/L$, the following results are obtained.

$$e^{-Ls} = 1\angle - 57.296°, \tag{3.89}$$

$$1/A_L(s) = 0.97129\angle - 60.945°. \tag{3.90}$$

Generally speaking, the closed-loop system with time delay, must contain an integrator (or slow time-lag) for stability reasons. Its phase lag is 90°. For the closed-loop system to keep stable operation, the phase margin of at least 40° is necessary. Thus, the phase lag allowed to the time delay is less than 50°. This is the justification of the second reason. The third reason is that the stability condition for the closed-loop system with an integrator and a time delay is closely approximated. The open-loop transfer function $G(s)$ is as follows:

$$G(s) = e^{-Ls}K/s \simeq K/(sA_L(s)). \tag{3.91}$$

For the exact time delay, the oscillation frequency ω at stability limit and stability condition are obtained by the condition that the loop-gain is unity and phase lag is $-\pi$ rad $= -180°$.

$$\omega = \pi/(2L) = 1.5708/L, \quad K = \omega = 1.5708/L. \tag{3.92}$$

For the approximated time delay, the oscillation frequency and stability condition are obtained by the condition $P(j\omega) = 0$, where $P(s)$ is the characteristic polynomial of the closed-loop system.

$$P(s) = sA_L(s) + K = 0.1L^3s^4 + 0.5L^2s^3 + Ls^2 + s + K, \tag{3.93}$$

$$P(j\omega) = (0.1L^3\omega^4 - L\omega^2 + K) + j\omega(-0.5L^2\omega^2 + 1) = 0, \tag{3.94}$$

$$\omega = 1.4142/L, \quad K = 1.6/L. \tag{3.95}$$

Thus, the value of K for the stability condition is approximated with sufficient accuracy. The last fourth reason is that step response is approximated with sufficient accuracy. It is known that, at the system consisting of an integrator with proper gain and a time delay, the step response $W(s)$ is an idealistic response with no-overshoot [4]. The $W(s)$ for this case is shown as follows:

$$W(s) = 0.4/(Lse^{Ls} + 0.4). \tag{3.96}$$

For approximated case, it is shown as follows:

$$W(s) = 0.4/(0.1L^4s^4 + 0.5L^3s^3 + L^2s^2 + Ls + 0.4). \tag{3.97}$$

For this case, the stability indexes $[\gamma_3\ \gamma_2\ \gamma_1] = [2.5\ 2\ 2.5]$ are large and there is no overshoot. The comparison of the responses are shown in Fig. 3.22 for $L = 1$. The approximation is sufficiently accurate.

Approximation of Discrete-Time System is next discussed. The input is considered to be a time function $x(t) = t$. This input is sampled at the sampling time of every T_s and is sent out as the output after T_s. The output is $y(t) = x(t - T_s)$, and the output is kept at this value by the sample-hold device until the next sampling time. The average value is further delayed by $0.5T_s$. This relation is shown in Fig. 3.23. Thus, it is concluded that, when the continuous signal goes through a discrete-time system, the output $y_{out}(t)$ is delayed by $1.5T_s$. The relation is shown as follows:

$$y_{out}(t) = x(t - 1.5T_s) = e^{-1.5T_s s}x(t). \tag{3.98}$$

Thus, the design of the controller of a discrete-time system can be made with this time delay in consideration. In this controller, "Digital-Out and Digital-In" is repeated at the sampling time. Because the time delay is fixed at $1.5T_s$ in such systems, proper compensation can be easily made.

As explained above, the discrete-time system can be handled as the continuous system by addition of the time delay. However precaution is necessary, as the value of such time delay changes according to the situation. The first example is the case where the input is $x(t) = t$ and the output is its integral with the Euler approximation formula. The output is $y(nT_s) = y((n-1)T_s) + T_s x((n-1)T_s)$.

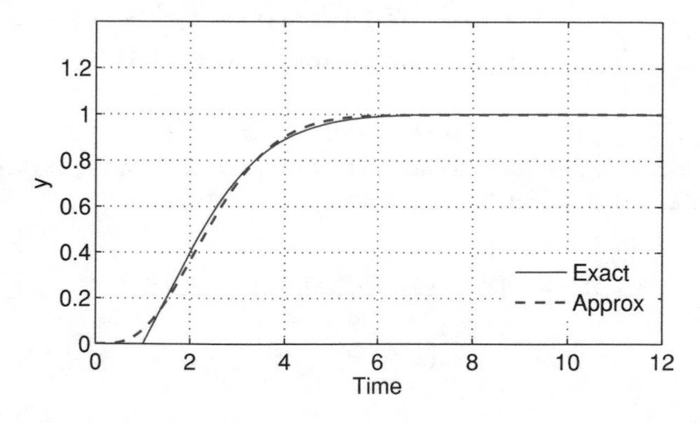

Fig. 3.22 Comparison of responses

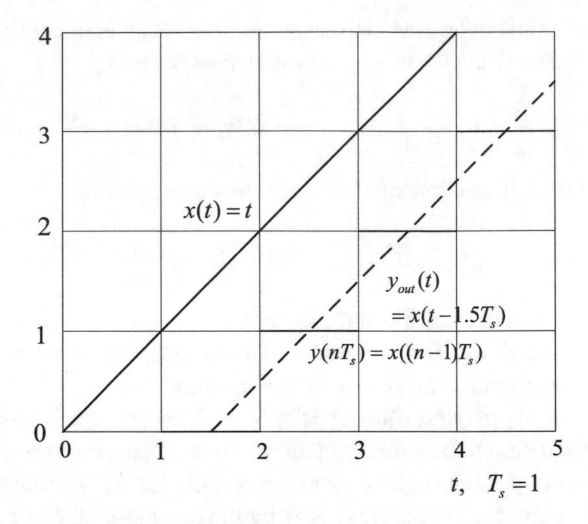

Fig. 3.23 Discrete-time system

The relation $[y(nT_s) - y((n-1)T_s)]/T_s = sy((n-0.5)T_s)$ holds approximately. Then, the input–output is expressed as $sy((n-0.5)T_s) = x((n-1)T_s) = x((n-0.5)T_s - 0.5T_s)$. For $t = (n-0.5)T_s$, the input–output relation is expressed as $y(t) = (1/s)x(t-0.5T_s)$. Because the output is kept at this value until next sampling time, further time delay of $0.5T_s$ is added. As the result, the output is the integral of the input with time delay of one sample time as shown below.

$$y_{out}(t) = \frac{1}{s}x(t-T_s) = \frac{e^{-T_s s}}{s}x(t). \tag{3.99}$$

The second case is when Advancing-Modified-Euler approximation formula is used. The output is expressed as $y(nT_s) = y((n-1)T_s) + 1.5T_s x((n-1)T_s) - 0.5T_s x((n-2)T_s)$. This equation is further modified as follows:

$$\begin{aligned} sy((n-0.5)T_s) &= 1.5x((n-1)T_s) - 0.5x((n-2)T_s) \\ &= x((n-1)T_s) + 0.5x((n-1)T_s - x((n-2)T_s) \\ &= x((n-1)T_s) + 0.5T_s sx((n-1.5)T_s). \end{aligned} \tag{3.100}$$

For $(n-0.5)T_s = t$, this equation becomes as $y(t) = (1/s)[x(t-0.5T_s) + 0.5T_s sx(t-T_s)]$. Because this output is kept at this value until next sampling time, further time delay of $0.5T_s$ is added. Then, the output is as follows:

$$\begin{aligned} y_{out}(t) &= \frac{1}{s}[x(t-T_s) + 0.5T_s sx(t-1.5T_s)] \\ &= \frac{e^{-0.5T_s s}}{s}[e^{-0.5T_s s}x(t) + 0.5T_s e^{-T_s s}sx(t)]. \end{aligned} \tag{3.101}$$

For $x(t) = t$, this is expressed as follows:

$$y_{out}(t) = \frac{e^{-0.5T_s s}}{s}x(t) = \frac{1}{s}x(t - 0.5T_s). \tag{3.102}$$

In this case, the integration is accurately made. However, the time delay of $0.5T_s$ is introduced, because the output is kept at this value until the next sampling time. As explained above, the introduction of a discrete-time system corresponds to the introduction of time delay of $(0.5–1.5)T_s$. The total design is to be made with this in consideration. The total time delay of the system is the sum of the original time delay L and that due to sampling $(0.5–1.5)T_s$. Thus, the time delay due to sampling is usually neglected as $T_s \ll L$.

Now **Design of PID Control** is made by CDM. The block diagram of the control system is the standard block diagram for CDM as shown in Fig. 3.5. The plant consists of a time delay and a time-lag as follows:

$$\begin{aligned}
G_p(s) =& \frac{B_p(s)}{A_p(s)} = \frac{e^{-Ls}K}{Ts+1} \\
\simeq& \frac{K}{A_L(s)(Ts+1)} = \frac{R}{A_L(s)(s+1/T)}, \quad R = K/T, \\
A_L(s) =& 0.1L^3s^3 + 0.5L^2s^2 + Ls + 1, \\
A_p(s) =& 0.1L^3s^4 + 0.5L^2(1+0.2L/T)s^3 \\
&+ L(1+0.5L/T)s^2 + (1+L/T)s + 1/T, \\
B_p(s) =& R.
\end{aligned} \tag{3.103}$$

As the controller, the simple PI controller (1/1 order controller) is assumed. This assumption is found to be valid after the design is complete.

$$G_c(s) = \frac{B_c(s)}{A_c(s)} = \frac{k_1 s + k_0}{s}. \tag{3.104}$$

The characteristic polynomial is as follows:

$$P(s) = A_c(s)A_p(s) + B_c(s)B_p(s) = \sum_{i=0}^{5} a_i s^i, \tag{3.105}$$

where

$$\begin{aligned}
&a_5 = 0.1L^3, \quad a_4 = 0.5L^2(1+0.2L/T), \quad a_3 = L(1+0.5L/T), \\
&a_2 = 1 + L/T, \quad a_1 = 1/T + Rk_1, \quad a_0 = Rk_0.
\end{aligned} \tag{3.106}$$

Design is made with stability indexes of $\gamma_2 = 2$ and $\gamma_1 = 2.5$. The second-order equivalent time constant τ_2 is obtained from the parameters of the plant. Then, the first-order equivalent time constant τ_1 and the equivalent time constant τ are obtained as follows:

$$\tau_2 = \frac{a_3}{a_2} = \frac{L(1+0.5L/T)}{1+L/T}, \qquad \tau_1 = \gamma_2 \tau_2, \tag{3.107}$$

$$\tau = \gamma_2 \gamma_1 \tau_2 = \frac{5L(1+0.5L/T)}{1+L/T}. \tag{3.108}$$

If $L/T = 0$, then $\tau_2 = L, \tau_1 = 2L, and \tau = 5L$. The controller parameters are obtained as follows:

$$a_1 = \frac{a_2}{\tau_1} = \frac{0.5(1+L/T)^2}{L(1+0.5L/T)}, \tag{3.109}$$

$$a_0 = \frac{a_1}{\tau} = \frac{0.1(1+L/T)^3}{L^2(1+0.5L/T)^2}, \tag{3.110}$$

$$k_1 = \frac{a_1 - 1/T}{R} = \frac{0.5}{RL(1+0.5L/T)}, \tag{3.111}$$

$$k_0 = \frac{a_0}{R} = \frac{0.1(1+L/T)^3}{RL^2(1+0.5L/T)^2}. \tag{3.112}$$

Because in ordinary plant $L/T = 0$ is approximately satisfied, the case for $L/T = 0$ is shown as an example. The results become as follows:

$$a_1 = 0.5/L, \quad a_0 = 0.1/L^2, \tag{3.113}$$

$$k_1 = 0.5/(RL), \quad k_0 = 0.1(RL^2). \tag{3.114}$$

Also, the design can be simply made by systematically filling the data in the CDM design form as shown in Fig. 3.24.

The above design can be made in much simpler manner with CDM Toolbox. In this case, equivalent time constant τ is obtained by Eq. (3.107) and command gc is used. The controller parameters are automatically calculated from the equivalent time constant and stability indexes. The command line for $R = L = 1$ and $1/T = 0$ is as follows:

```
>> RR=1;L=1;Tinv=0;aL=[0.1*L^3 0.5*L^2 L 1];
>> ap=[conv(aL [1 Tinv]) 0];bp=RR;nc=0;mc=1;
>> gr=[2 2 2 2.5];t=5*L*(1+0.5*Tinv)/(1+L*Tinv);
>> tm=0.5; gc
```

In the above, the integrator of the controller is moved to the plant. Thus, the order of denominator of the controller n_c is 0, while that of numerator m_c is 1. The gr is the reference stability index. Because the freedom of design is two, k_1 and k_0, only γ_1 and γ_2 take the reference values. Other stability indexes, γ_3 and γ_4, are obtained as the design results, which are to be accepted as they are. The t_m is used to change the time scale of the step response. Usually, it is 0.5 or 1. For the case $R = L = 1$ and $1/T = 0$, the summary of design results is as follows:

$$G(s) = G_c(s)G_p(s) = \frac{0.5s + 0.1}{s} \frac{1}{0.1s^4 + 0.5s^3 + s^2 + s}, \tag{3.115}$$

$$P(s) = 0.1s^5 + 0.5s^4 + s^3 + s^2 + 0.5s + 0.1,$$

$$\gamma_i = [2.5\ \ 2\ \ 2\ \ 2.5], \quad \tau = 5,$$

$$s_i = -1, \ -1, \ -1, \ -1,$$

$$\phi = 38.319°, \quad g_m = 2.7778.$$

The coefficient diagram is shown in Fig. 3.24, and the design results are shown in Fig. 3.25.

Finally, **Comparison with Ziegler–Nichols Approaches** is made with these CDM results. In these comparisons, the time constant of the time-lag of the plant is considered very large compared with its time delay, namely $L/T = 0$. The transfer function of the plant is $G_p(s) = e^{-Ls}R/s$. As the controller, P control, PI control, and PID control are considered. However, because PI control is considered already, and PID control is found to be unnecessary, only P control is considered here. The open-loop transfer function $G(s)$ and the characteristic polynomial $P(s)$ are shown below.

$$G(s) = G_c(s)G_p(s) = k_0 e^{-Ls} R/s = k_0 R/(sA_L(s)), \tag{3.116}$$

$$P(s) = 0.1L^3s^4 + 0.5L^2s^3 + Ls^2 + s + Rk_0. \tag{3.117}$$

In CDM design, $\gamma_1 = 1/(LRk_0) = 2.5$. Then $k_0 = 0.4/(RL)$. In PI control, design results are shown in Eq. (3.113). In ordinary PI control, the controller is expressed as follows:

$$G_c(s) = K_P\left(1 + \frac{1}{T_I s}\right). \tag{3.118}$$

With this expression, the design results by CDM are as follows:

$$K_P = 0.4/(RL), \qquad \mathit{for\ P\ control}, \tag{3.119}$$

$$K_P = 0.5/(RL), \quad T_I = 5L, \quad \mathit{for\ PI\ control}. \tag{3.120}$$

Two methods are proposed in Ziegler–Nichols approach [1, p.191]. The first method is called Quarter Decay Ratio (QDR). In this method, controller parameters are selected in such a manner that the amplitude decay at one cycle of oscillation is

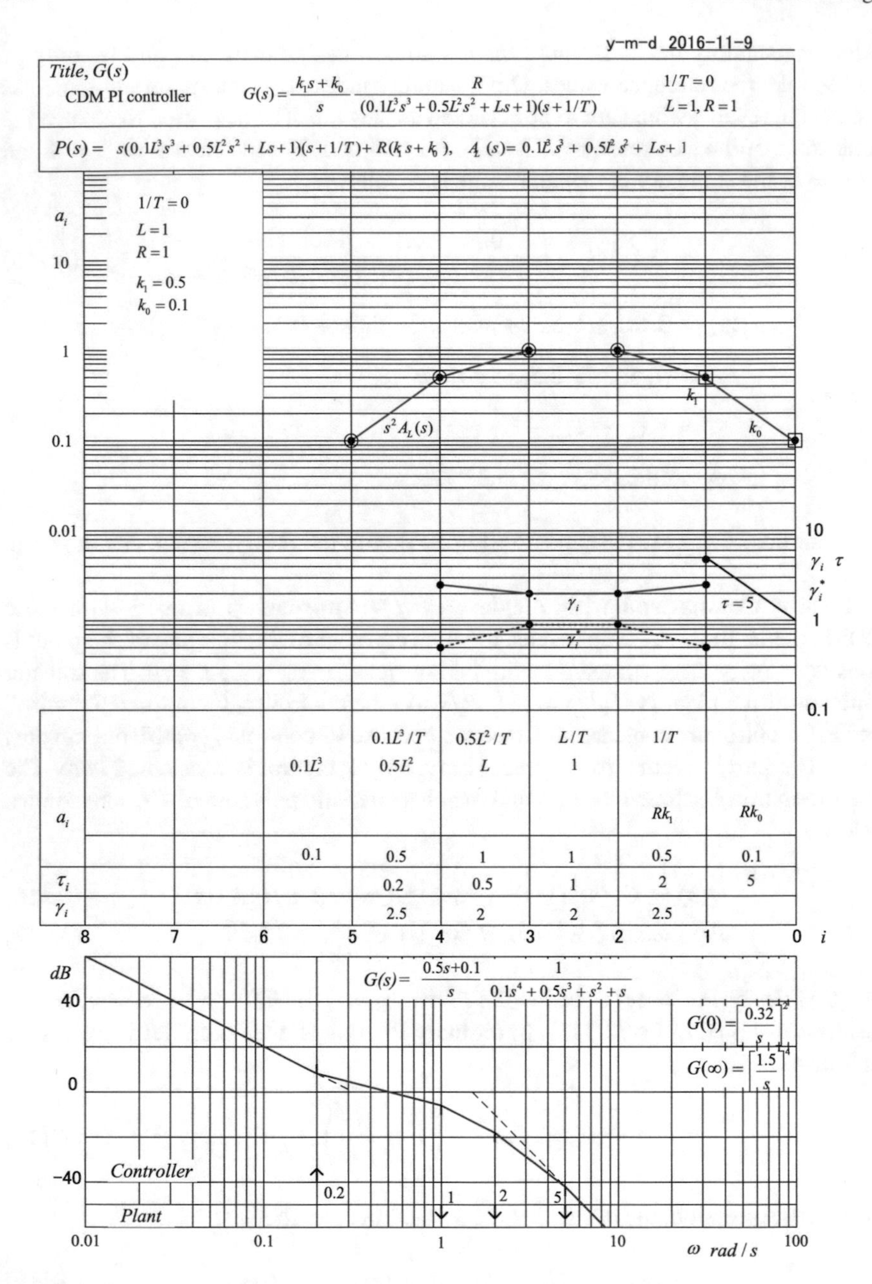

Fig. 3.24 Coefficient diagram, CDM, R=L=1, 1/T=0

```
>>RR=1;L=1;Tinv=0;aL=[0.1*L^3 0.5*L^2 L 1];ap=[conv(aL,[1 Tinv]) 0];bp=RR;
>> nc=0;mc=1; gr=[2 2 2 2.5];t=5*L*(1+0.5*Tinv)/(1+L*Tinv);tm=0.5, gc

tm =
   0.5000
ba =
   0.1000
bc =
   0.5000   0.1000
ac =
     1
aa =
   0.1000   0.5000   1.0000   1.0000   0.5000   0.1000
g =
   2.5000   2.0000   2.0000   2.5000
tau =
     5
gs =
   0.5000   0.9000   0.9000   0.5000
rr =
  -1.0009 + 0.0007i
  -1.0009 - 0.0007i
  -0.9997 + 0.0011i
  -0.9997 - 0.0011i
  -0.9989 + 0.0000i
pmgm =
  38.3190   2.7778
wpmgm =
   0.5329   1.2910
```

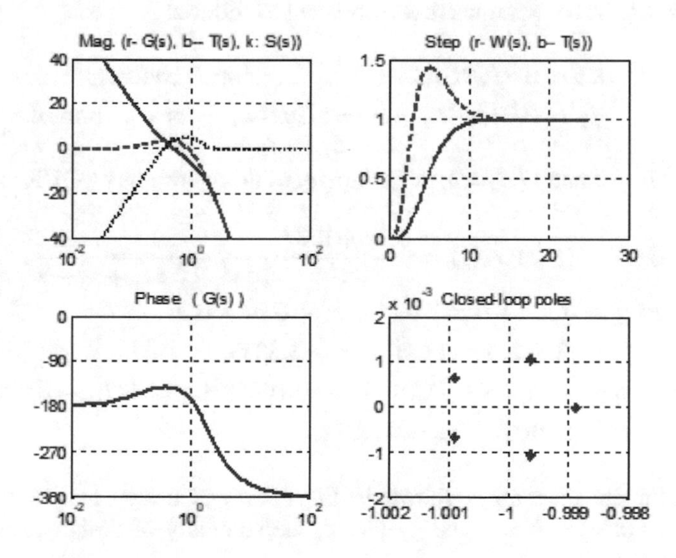

Fig. 3.25 Design results, CDM, R=L=1, 1/T=0

about 0.25. In this case, the controller parameters are as follows:

$$K_P = 1/(RL), \qquad for\ P\ control, \tag{3.121}$$

$$K_P = 0.9/(RL), \quad T_I = 3.3333L, \quad for\ PI\ control. \tag{3.122}$$

The second method is called a Marginally Stable System (MSS). The controller parameters are selected on the basis of gain limit K_U and oscillation period P_U at the stability limit in P control as shown below.

$$K_P = 0.5K_U, \qquad for\ P\ control, \tag{3.123}$$

$$K_P = 0.45K_U, \quad T_I = P_U/1.2, \quad for\ PI\ control. \tag{3.124}$$

The characteristic polynomial is as follows:

$$P(s) = 0.1L^3s^4 + 0.5L^2s^3 + Ls^2 + s + RK_U. \tag{3.125}$$

From the stability condition for the fourth-order system, the K_U and P_U are obtained as follows:

$$K_U = 1.6/(RL), \qquad P_U = 2\pi/(1/\sqrt{0.5L^2}) = \sqrt{2}\pi L. \tag{3.126}$$

Then, the controller parameters are obtained as follows:

$$K_P = 0.8/(RL), \qquad for\ P\ control, \tag{3.127}$$

$$K_P = 0.72/(RL), \quad T_I = 3.7024L, \quad for\ PI\ control. \tag{3.128}$$

For $R = L = 1$ and $1/T = 0$, the summary of design results by QDR is as follows:

$$G(s) = G_c(s)G_p(s) = \frac{0.9s + 0.27}{s}\frac{1}{0.1s^4 + 0.5s^3 + s^2 + s}, \tag{3.129}$$

$$\begin{aligned} P(s) &= 0.1s^5 + 0.5s^4 + s^3 + s^2 + 0.9s + 0.27, \\ \gamma_i &= [2.5\ \ 2\ \ 1.1111\ \ 3], \quad \tau = 3.3333, \\ s_i &= -2.1205 \pm j1.0154, \ -0.16008 \pm j1.0427, \ -0.4389, \\ \phi &= 16.046^\circ, \quad g_m = 1.4084. \end{aligned}$$

The coefficient diagram is shown in Fig. 3.26, and design results are shown in Fig. 3.27. For $R = L = 1$ and $1/T = 0$, the summary of design results by MSS is as follows:

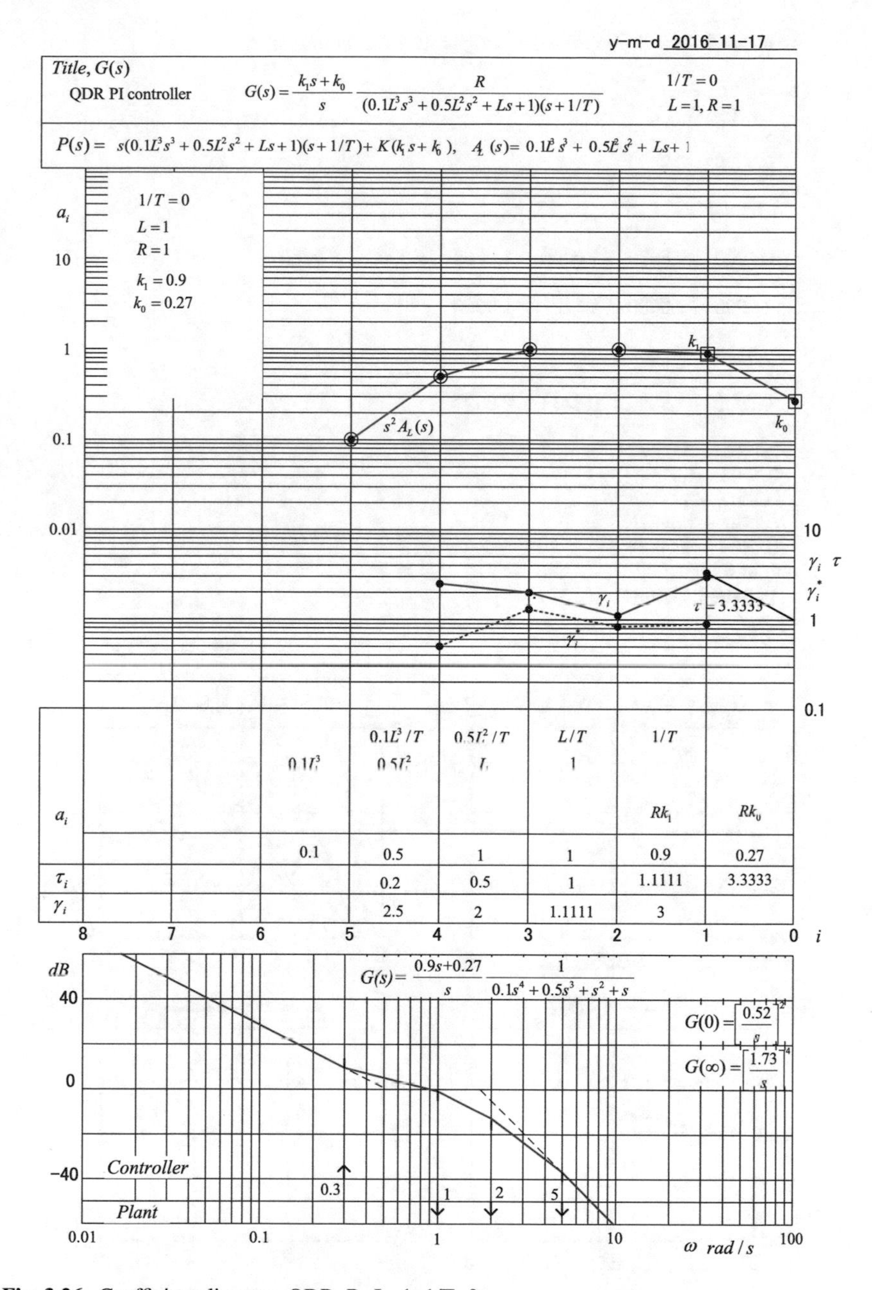

Fig. 3.26 Coefficient diagram, QDR, R=L=1, 1/T=0

```
>>ap=[0.1 0.5 1 1 0];bp=[1];ac=[1 0];bc=[0.9 0.27];tm=0.5; cc

ba =
   0.2700
bc =
   0.9000    0.2700
ac =
    1     0
aa =
   0.1000    0.5000    1.0000    1.0000    0.9000    0.2700
g =
   2.5000    2.0000    1.1111    3.0000
tau =
   3.3333
gs =
   0.5000    1.3000    0.8333    0.9000
rr =
  -2.1205 + 1.0154i
  -2.1205 - 1.0154i
  -0.1601 + 1.0427i
  -0.1601 - 1.0427i
  -0.4389 + 0.0000i
pmgm =
  16.0459    1.4084
wpmgm =
   0.9263    1.2204
```

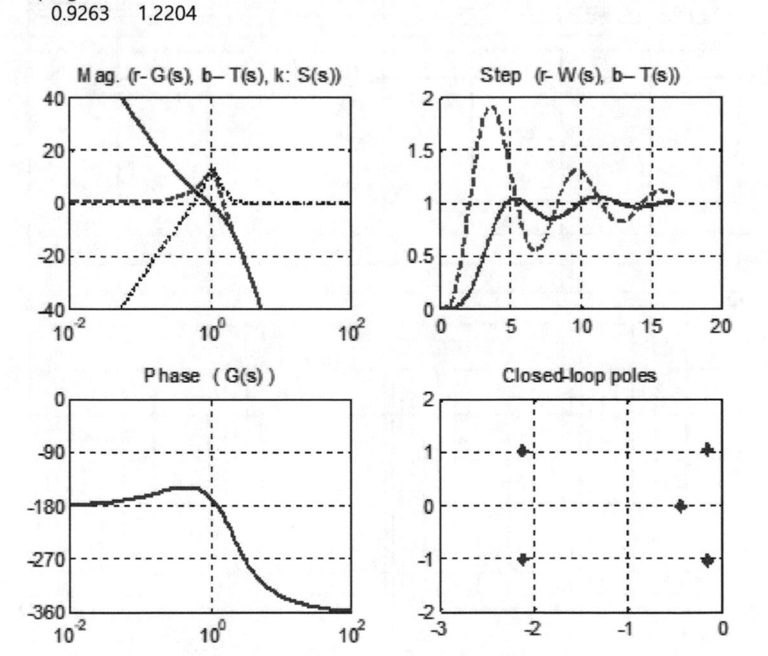

Fig. 3.27 Design results, QDR, R=L=1, 1/T=0

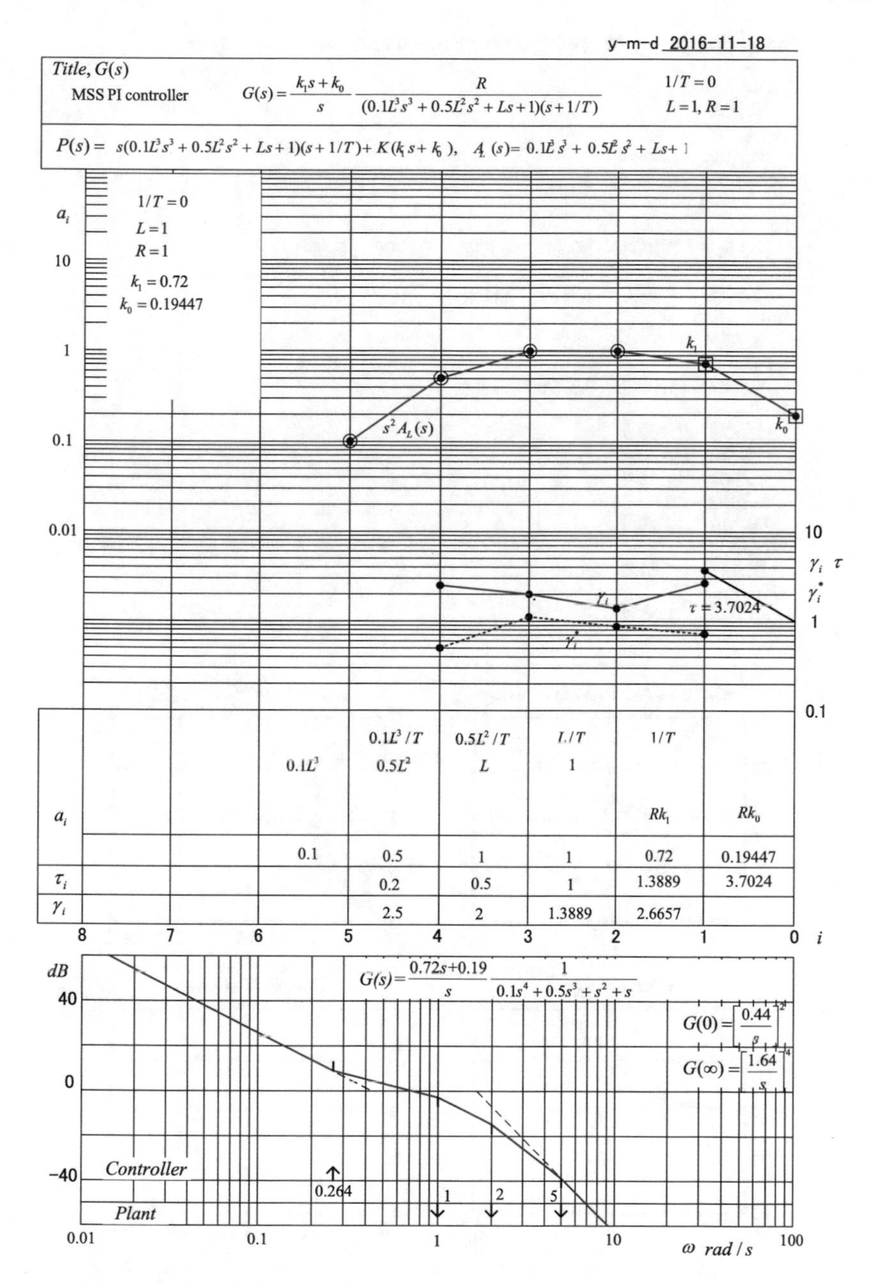

Fig. 3.28 Coefficient diagram, MSS, R=L=1, 1/T=0

```
>>ap=[0.1 0.5 1 1 0]; bp=[1]; ac=[1 0]; bc=[0.72 0.19447]; tm=0.5;  cc

ba =
   0.1945
bc =
   0.7200    0.1945
ac =
   1    0
aa =
   0.1000    0.5000    1.0000    1.0000    0.7200    0.1945
g =
   2.5000    2.0000    1.3889    2.6657
tau =
   3.7024
gs =
   0.5000    1.1200    0.8751    0.7200
rr =
  -1.9778 + 0.8776i
  -1.9778 - 0.8776i
  -0.2962 + 0.9118i
  -0.2962 - 0.9118i
  -0.4519 + 0.0000i
pmgm =
  25.3374    1.8125
wpmgm =
   0.7575    1.2422
```

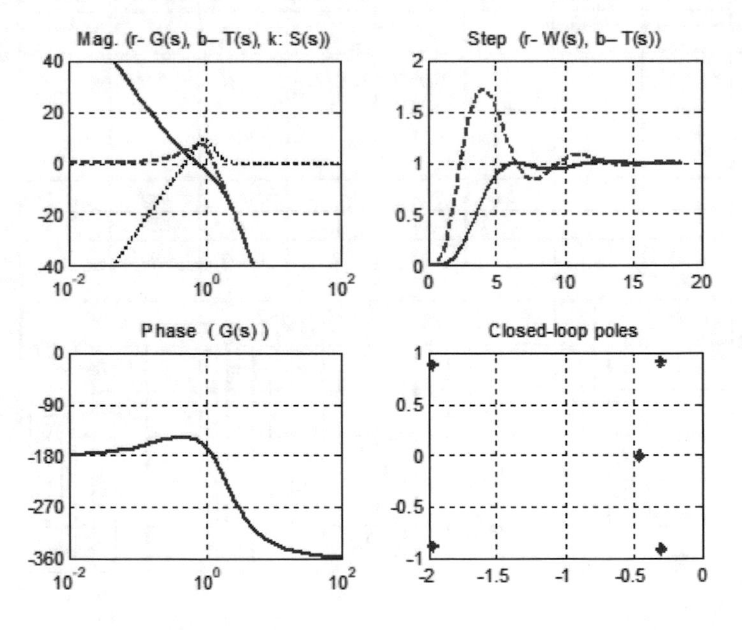

Fig. 3.29 Design results, MSS, R=L=1, 1/T=0

$$G(s) = G_c(s)G_p(s) = \frac{0.72s + 0.19447}{s} \frac{1}{0.1s^4 + 0.5s^3 + s^2 + s}, \quad (3.130)$$
$$P(s) = 0.1s^5 + 0.5s^4 + s^3 + s^2 + 0.72s + 0.19447,$$
$$\gamma_i = [2.5\ 2\ 1.3889\ 2.6657], \quad \tau = 3.7024,$$
$$s_i = -1.9778 \pm j0.87761, \ -0.29624 \pm j0.91181, \ -0.45189,$$
$$\phi = 25.337°, \quad g_m = 1.8125.$$

The coefficient diagram is shown in Fig. 3.28, and design results are shown in Fig. 3.29.

The step responses of P control are shown in Fig. 3.30, where $R = L = 1$ and $1/T = 0$. The K_Ps are 0.4 (CDM), 1(QDR), and 0.8(MSS), respectively. The step response of QDR shows the amplitude decay of about 0.25 in one cycle of oscillation. The values of K_P are 2.5 times in QDR and 2 times in MSS compared with CDM. The systems are more oscillatory. The step responses of complementary sensitivity function $T(s)$ for PI control are shown in Fig. 3.31. The tendency is similar to P control. Because the settling time is shortest in CDM, there is no need in increasing gain as in QDR and MSS.

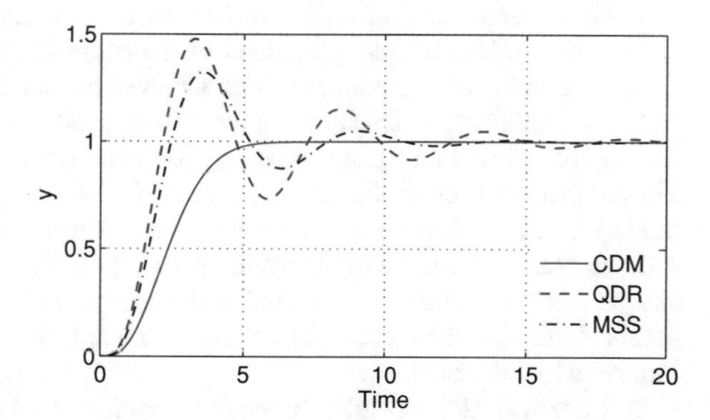

Fig. 3.30 Step responses with P controllers

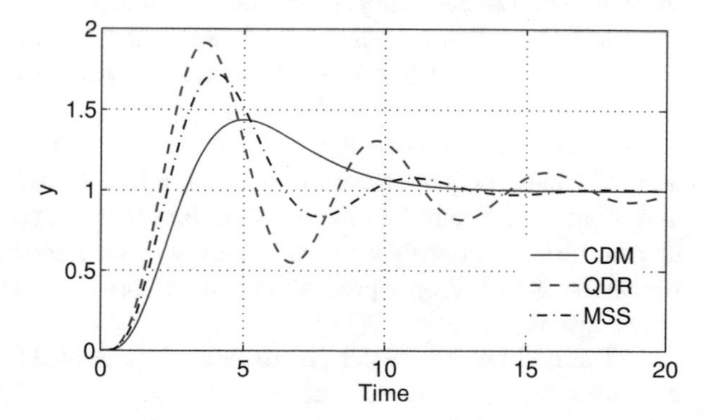

Fig. 3.31 Step responses with PI controllers

3.8 Summary

The important points discussed in this chapter are summarized in the following. In **Interpretation of Design Specification**, the design specification used in CDM is compared with those used in current control design approaches. In CDM, soft design specifications are given at first, and they are converted to more concrete specifications as the design proceeds, in a trial-and-error manner. Thus, it is difficult to relate them to those used in Pole Placement, Time Response, Sensitivity/Complementary Sensitivity Functions, and Open-loop Transfer Function approaches. However, the design results by CDM coincide with the specifications recommended in such design approaches. In short, the design specifications in CDM are summarized to the following three points; namely, the stability indexes should be chosen in the values close to the standard form, the controller parameters are to be chosen as non-negative, and the equivalent time constant should be chosen in trade-off consideration between the controller complexity and the value of the equivalent time constant.

In **Definition of Basic Control Structure**, the process of controller design is explained, where due attention is paid to the order of controller and values of controller parameters. In this process, disturbance rejection characteristics and noise attenuation characteristics become important. In disturbance rejection characteristics, only the removal of constant disturbance is usually required. This can be achieved by making the lowest order parameter of the controller denominator, l_0, to 0. The noise attenuation characteristics can be achieved by increasing the order of the controller denominator, n_c. Because the order of controller is greatly influenced by "the easiness of control of the plant", it is very difficult to make the general design theory. The recommended controllers are 1/1 order (PI), 2/2 order (generalized PID), and 3/3 order, where $l_0 = 0$. Because "the easiness of control of the plant" does not necessarily depend on the order of the plant, design starts with the lowest order controller, and the controller order is increased until the required equivalent time constant is obtained. If noise attenuation characteristics are not sufficient at this point, it can be improved by increasing n_c.

In **Interpretation of Robustness**, the meaning of robustness is made clear, and the consideration necessary for attaining such robustness is clarified. In CDM, only the case where robustness is weak while stability is sufficient is discussed. Because the cause for this situation is that the controller parameter is negative, robustness is attained by making the controller parameters non-negative. However, when the plant is close to pole-zero cancellation (weak controllability or observability), stability and robustness become trade-off relation, and control becomes difficult. Stability and robustness cannot be satisfied simultaneously. Also, when the equivalent time constant is not cautiously selected, some controller parameter becomes negative, and robustness is lost. Proper precaution is necessary for the selection of the equivalent time constant.

In **Design Process**, the design procedure is presented for the ordinary plant, where controller design is not difficult. For such a plant, design proceeds on the proper selection of stability indexes and equivalent time constant with a non-negative selection of controller parameters. A shorter equivalent time constant will be obtained by increasing the controller order from 1/1 to 2/2, etc. In this design process, the stability indexes become the CDM standard except for the highest order one. Thus, the most proper equivalent time constant is obtained as the design result. This is the most conspicuous feature of CDM, "Simultaneous design".

In **Position Control**, the control design of simple position control with velocity feedback is shown. The design is made with the CDM standard form. The equivalent time constant is automatically obtained as the design result. The method to draw the CDM-type Bode diagram is shown.

In **Position Control with Integrator**, the position control with the integrator for disturbance compensation is discussed. Because the plant is a real toy model, various problems which arise in real control design are also considered. First, the method of building the mathematical model is shown. The saturation due to current limitation is discussed. At the current saturation, the "integrator wind-up" takes place. The saturation of the integrator to prevent integrator wind-up is discussed. Also, some measures are taken for the design of the controller to guarantee a smooth transition from nonlinear operation with saturation to linear operation without saturation. Such measures are briefly explained.

In **PID Control**, the design by CDM is first presented. The results are compared with those by the well-known Ziegler–Nichols. The plant in consideration is approximated by a time delay and a time-lag. The time delay is approximated by a third-order denominator polynomial, so that it can be easily handled in CDM. Also, the discrete-time system usually found in ordinary PID control is approximated by the continuous system with equivalent time delay such that CDM is applicable. In CDM design, there is no overshoot in its step response and the settling time is short. In Ziegler–Nichols design, the gains are about 2–2.5 times large. Their step responses are fast with short response times. However, the responses are oscillatory and the settling times are longer.

References

1. Franklin CF, Powell JD, Emami-Naeini A (1994) Feedback control of dynamic systems, 3rd edn. Addison-Wesley, Boston
2. Franklin CF, Powell JD, Emami-Naeini A (2015) Feedback control of dynamic systems, 7th edn. Pearson Education Ltd., Essex
3. Chen CT (1987) Introduction to the linear algebraic method for control system design. IEEE Control Syst Mag 7(5):36–42
4. Manabe S (2003) Early development of fractional order control. In: Proceedings of the ASME 2003 design engineering technical conferences, Chicago, Illinois, USA, September 2–6, 2003, pp 609–616

Chapter 4
Advanced Controller Design: Case Studies

Abstract In this chapter, four selected examples of applying CDM to advanced practical control design problems are presented, **A low-cost inverted pendulum control** is the case where CDM is applied to an unstable plant. Contrary to ordinary control of the inverted pendulum where both pendulum angle and cart position are required to be sensed, only pendulum angle is needed in this CDM design. **Vibration suppression control for two-inertia system** is very important but not easy to design. CDM designs of PID and first-order controllers for this system are shown. The third example is a **CDM solution to the American Control Conference (ACC) benchmark problem**, where two masses are connected by a spring and a control force is applied to one mass while the position of the other mass is measured. Also, CDM is applied to a multi-input multi-output (MIMO) control problem, which is to design **a longitudinal controller for a modern aircraft**. The results are compared with the performance of H_∞ controller. Although not introduced here, there are other results of applying CDM to well-known control problems, for example, acceleration control of a dual-control surface missile (Manabe in Proceedings of the 15th IFAC symposium on automatic control in aerospace, Bologna, Italy, Sept 2–7, pp 499–504, 2001 [1]), hot rolling mill control (Kim et al. in Trans Control Autom Syst Eng 3(4):2091–2096, 2001 [2]), and control of an active suspension system (Kim et al in Proceedings of the 4th IFAC symposium on robust control design, Milan, Italy, vol 36, no 11, pp 55–60, 2003 [3]), which was proposed as the benchmark problem of European Journal Control.

4.1 A Low-Cost Inverted Pendulum System

The inverted pendulum system is a well-known balancing control problem. This system consists of an inverted pendulum mounted on a cart. Two-wheeled robots such as Segway®, and the problem of controlling the attitude of a missile during the initial stages of the launch are practical examples of such systems. In this section, we introduce a low-cost inverted pendulum system that can be implemented for only $200. The system is composed of a battery-powered toy car, a small pendulum, a magnetic type potentiometer for non-contact angle detection, and an analog controller. For

S. Manabe and Y. C. Kim, *Coefficient Diagram Method for Control System Design*, Intelligent Systems, Control and Automation: Science and Engineering 99,
https://doi.org/10.1007/978-981-16-0546-8_4

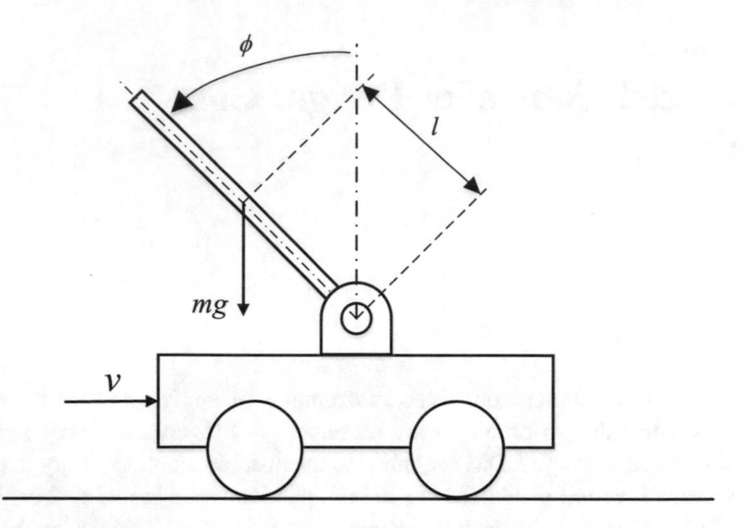

Fig. 4.1 Inverted pendulum

details, refer to Manabe [4], and here we will focus on the design of a balancing controller using CDM.

Consider the inverted pendulum in Fig. 4.1. This model is for a car with a mass of 0.25 Kg, a pendulum 0.2 m long and 0.048 Kg weight. Since the center of gravity (COG) of the pendulum is in the middle due to the uniform density of the rod, the distance from the pivot point to COG equals $l = 0.1$ m. After applying Newton's laws for the translational motion of the cart and the rotational motion of the pendulum, and eliminating the reaction forces between two bodies, the following linearized motion equation can be derived (for details, see Appendix W2.1.4 in Franklin's book [5]).

$$(I + ml^2)\ddot{\phi} - mgl\phi = Ml\dot{v}, \tag{4.1}$$

where for small motions about $\phi = 0$, $cos\phi \simeq 1$ and $sin\phi \simeq \phi$ were assumed, and friction was neglected. v denotes the velocity of the toy car and the inertia moment of rod about end is $I = \frac{1}{3}ml^2$. Suppose that the velocity of the cart is controlled by a velocity controller.

Simply rewriting Eq. (4.2) yields

$$\ddot{\phi} - a\,\phi = b\,\dot{v}, \tag{4.2}$$

where

$$a = \frac{mgl}{ml^2 + I} = 73.5, \qquad b = \frac{Ml}{ml^2 + I} = 39.06. \tag{4.3}$$

Now, the problem is to design a controller that achieves the following performance goals:

(1) The pendulum must be in an upright position even when there is an error in sensing the angle or the cart is running on a slope.

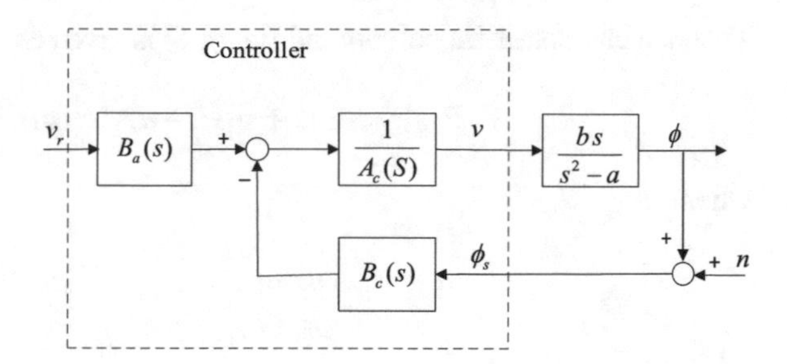

Fig. 4.2 Inverted pendulum control system

(2) The cart follows the reference velocity while the pendulum is kept in an upright position.
(3) The above performance is realized by utilizing only one angle sensor without a position sensor.

The control structure in two parameter configuration is considered, as shown in Fig. 4.2. v_r is the reference velocity and n is the measurement error of the angle sensor. The control law is

$$v(s) = \frac{1}{A_c(s)}[B_a(s)v_r(s) - B_c(s)\phi_s(s)], \tag{4.4}$$

The closed-loop system is given as

$$\begin{bmatrix} \phi \\ v \end{bmatrix} = \begin{bmatrix} \frac{b\ s}{P(s)} \\ \frac{(s^2-a)}{P(s)} \end{bmatrix} \lfloor B_a(s)\ v_r - B_c(s)\ n \rfloor + \begin{bmatrix} \frac{A_c(s)\ s}{P(s)} \\ \frac{-B_c(s)\ s}{P(s)} \end{bmatrix} \phi_o, \tag{4.5}$$

where $\phi_o = \phi(0)$ is the initial value and $P(s)$ is the characteristic polynomial

$$P(s) = (s^2 - a)\ A_c(s) + bs\ B_c(s). \tag{4.6}$$

The constraints of the controller to achieve $v = v_r$ in the steady state are as follows:

$$B_a(0) = \frac{-P(0)}{a}, \tag{4.7}$$

$$B_c(0) = 0. \tag{4.8}$$

The condition Eq. (4.8) is also necessary to eliminate the effect of the biased error of the angle sensor. With these in consideration, a controller of the following form is chosen here.

$$A_c(s) = s^2 + l_1\ s + l_o, \tag{4.9}$$

$$B_c(s) = k_2\ s + k_1\ s, \tag{4.10}$$

$$B_a(s) = m_0 \tag{4.11}$$

Then the characteristic polynomial Eq. (4.6) is given as

$$P(s) = a_4 s^4 + a_3 s^3 + a_2 s^2 + a_1 s + a_0, \tag{4.12}$$

where

$$\begin{aligned} a_4 &= 1 \\ a_3 &= l_1 + b\,k_2 \\ a_2 &= l_0 - a + b\,k_1 \\ a_1 &= -a\,l_1 \\ a_0 &= -a\,l_0 \end{aligned} \tag{4.13}$$

In CDM, the desired characteristic polynomial is generated by choosing the design parameters, $\{\gamma_i,\ \tau\}$. In this case, the fourth-order polynomial can be assigned by

$$P(s) = s^4 + \left(\frac{\gamma_1\gamma_2\gamma_3}{\tau}\right)s^3 + \left(\frac{\gamma_1^2\gamma_2^2\gamma_3}{\tau^2}\right)s^2 + \left(\frac{\gamma_1^3\gamma_2^2\gamma_3}{\tau^3}\right)s + \left(\frac{\gamma_1^3\gamma_2^2\gamma_3}{\tau^4}\right). \tag{4.14}$$

Then from Eqs. (4.13) and (4.14), the controller parameters can be obtained by solving the following algebraic equation:

$$\begin{bmatrix} b & 0 & 1 & 0 \\ 0 & b & 0 & 1 \\ 0 & 0 & -a & 0 \\ 0 & 0 & 0 & -a \end{bmatrix} \begin{bmatrix} k_2 \\ k_1 \\ l_1 \\ l_0 \end{bmatrix} = \begin{bmatrix} a_3 \\ a_2 + a \\ a_1 \\ a_0 \end{bmatrix} = \begin{bmatrix} \gamma_1\gamma_2\gamma_3/\tau \\ \gamma_1^2\gamma_2^2\gamma_3/\tau^2 + a \\ \gamma_1^3\gamma_2^2\gamma_3/\tau^3 \\ \gamma_1^3\gamma_2^2\gamma_3/\tau^4 \end{bmatrix}. \tag{4.15}$$

Also, from Eqs. (4.7) and (4.13) we have

$$m_0 = l_0. \tag{4.16}$$

According to the guide explained in Chap. 3, we choose $\gamma_1 = 2.5$ and $\gamma_2 = \gamma_3 = 4.0$. If the plant is unstable or weekly damped, it is more robust in stability to select the values of γ_2 and γ_3 larger than those of the CDM standard. The smaller τ, the faster the response speed, but it may lead to unnecessarily excessive control input and rapid transient response of ϕ. Here, $\tau = 1.5$ was chosen after a few trials. The resulting controller parameters and the characteristic polynomial coefficients are obtained as

$$[k_2\ k_1\ l_1\ l_0] = [0.7859\ \ 6.5015\ \ -4.0312\ \ -2.6875], \quad m_0 = -2.6875, \tag{4.17}$$

$$[a_4\ \ a_3\ \ a_2\ \ a_1\ \ a_0] = [1.0\ \ 26.667\ \ 177.78\ \ 296.30\ \ 197.53].$$

Using the CDM Matlab script `[g,tau,gs]=cdia(P)`, it provides the coefficient diagram for the above characteristic polynomial and the $\gamma - \gamma^*$ diagram, as shown in Fig. 4.3. It results in $[\gamma_1^*\ \ \gamma_2^*\ \ \gamma_3^*] = [0.25\ \ 0.65\ \ 0.25]$. Since the γ_i values are

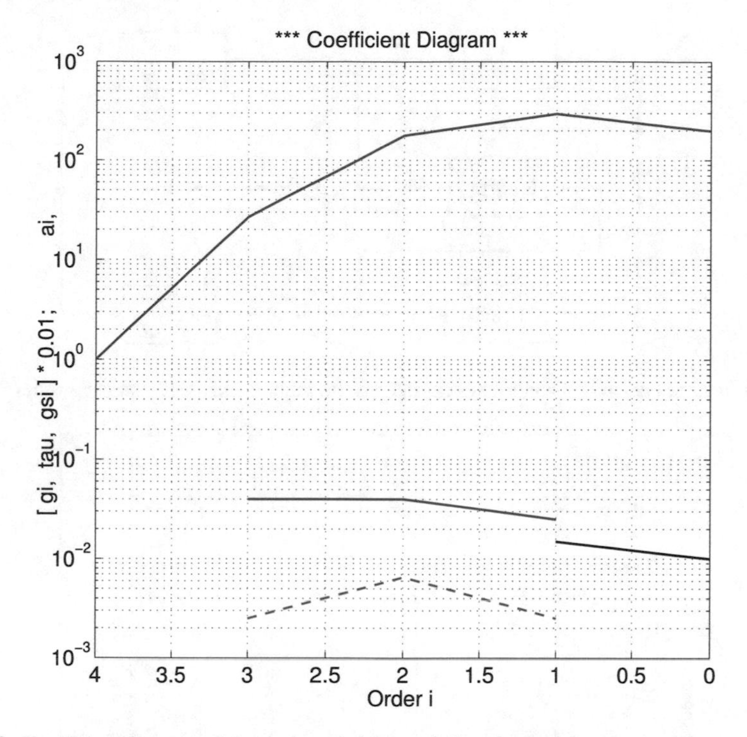

Fig. 4.3 Coefficient diagram of the designed characteristic polynomial

much greater than 1.5 times the γ_i^*, we can say that the designed controller has good robustness.

For an easier implementation of the controller Eq. (4.4), it has been transformed by mathematical manipulation as follows:

$$A_c(s) = s^2 - 4.0312s - 2.6875 = 2.6875(0.2167s - 1)(1.7167s + 1) \quad (4.18)$$

$$B_c(s) = 0.7859s^2 + 6.5015s = (0.458s + 3.521)(1.717s + 1) - 3.521 \quad (4.19)$$

$$B_a(s) = -2.6875 \quad (4.20)$$

$$v(s) = \frac{1}{(0.2167s - 1)}\left[\frac{(-v_r + 1.3099\,\phi_s)}{(1.7167s + 1)} - (0.1703s + 1.3099)\phi_s\right] \quad (4.21)$$

The overall system is shown in Fig. 4.4. When $v_r = 1$ and $\phi_o = 0$, the time responses v and ϕ of the closed-loop system are shown in Fig. 4.5, and when $v_r = 0$ and $\phi_o = 0.25\,\text{rad}$, the time responses from the initial value of the pendulum position is shown in Fig. 4.6. This inverter control system was successfully implemented in a real model and cost no more than \$200 [4].

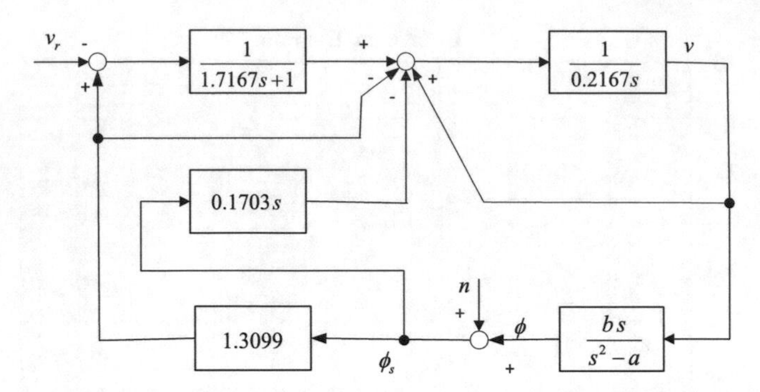

Fig. 4.4 Inverted pendulum control system that can be implemented with analog devices

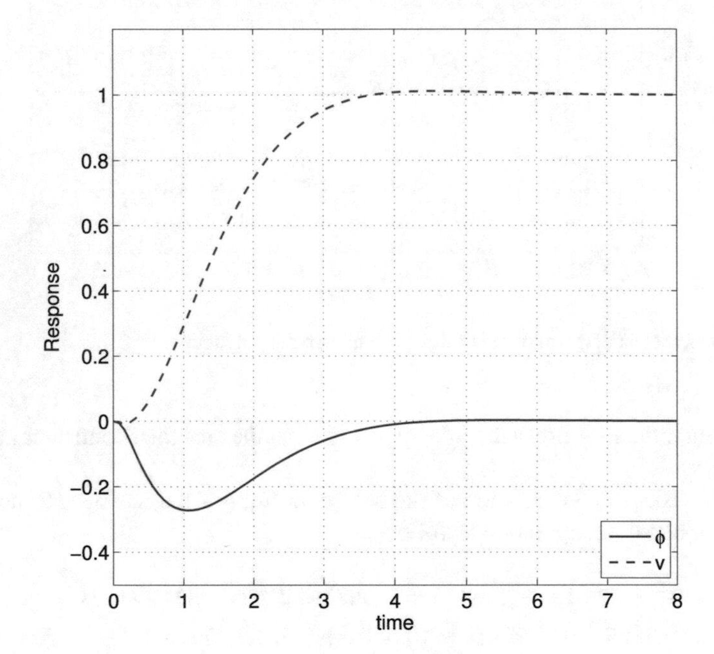

Fig. 4.5 Time response of the inverted control system when $v_r = 1$ and $\phi_o = 0$

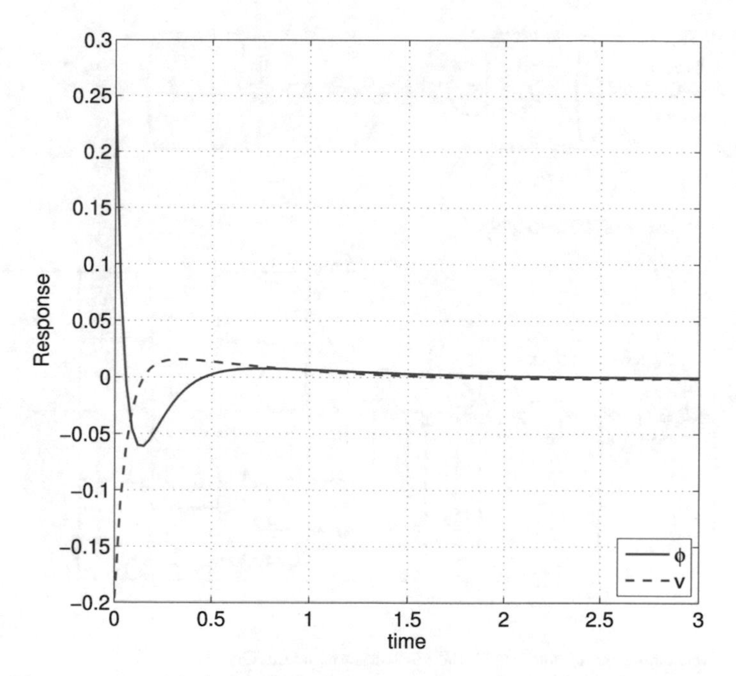

Fig. 4.6 Time response of the inverted control system when $v_r = 0$ and $\phi_o = 0.25$ rad

4.2 Vibration Suppression Control of Two-Inertia System

Two-inertia system is a simple model of actual engineering systems such as steel rolling mill, flexible robot arm, elevator, etc. Vibration suppression control for these systems is very important but not easy. CDM is applied to the design of a PID and first-order controllers for a two-inertia system shown in Fig. 4.7. Applying Newton's second law in each rotor, the following dynamic model equations are derived:

$$J_M \dot{\omega}_M = T_M - B_M \omega_M - K_s \theta_s \tag{4.22}$$

$$\dot{\theta}_s = \omega_M - \omega_L \tag{4.23}$$

$$J_L \dot{\omega}_L = -T_L - B_L \omega_L + K_s \theta_s, \tag{4.24}$$

where J_M and J_L denote the inertia moments of motor and load, J_M and J_L are their friction coefficients, and ω_M and ω_L are their rotational velocities, respectively. K_s is the spring constant.

Neglecting friction terms and introducing a disturbance observer by Ohnishi [6, 7], we consider the vibrational suppression control system, as shown in Fig. 4.8. The design purpose of the controller is to ensure that the step responses of motor and load velocities have satisfactory damping.

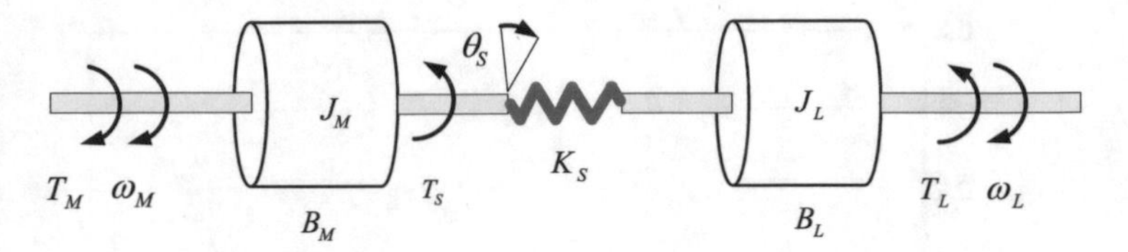

Fig. 4.7 Two-inertia system model

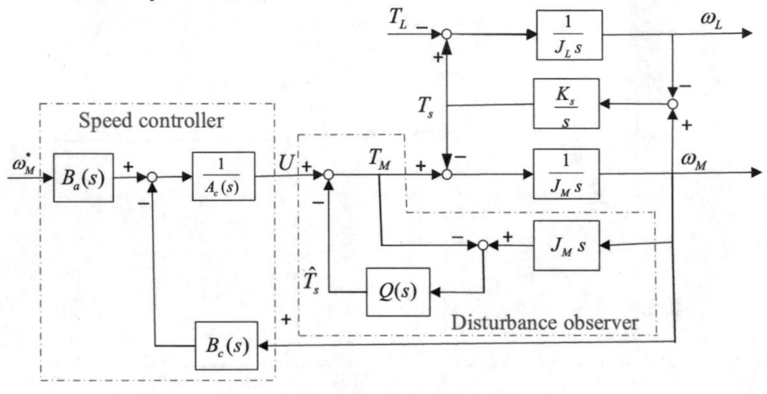

Fig. 4.8 Disturbance observer based control for two-inertia system

The equivalent plant transfer function from U to ω_M is given by

$$\frac{\omega_M}{U} = \frac{K_o(s^2 + \omega_a^2)}{s\,(s^2 + \omega_r^2)}, \quad \left(K_o = \frac{1}{J_M}\right), \tag{4.25}$$

where the resonant and anti-resonant frequencies are defined as

$$\omega_r := \sqrt{\frac{K_s}{J_L}\left(1 + \frac{J_L}{J_M}\right)}, \qquad \omega_a := \sqrt{\frac{K_s}{J_L}}. \tag{4.26}$$

It can be seen from these equations that the two-inertia system has a very weekly damped vibration mode. Design procedures of two vibration suppression controllers which are of I-PD and first-order types will be shown.

4.2.1 Design of an I-PD Controller

From Fig. 4.8, we first consider the following vibration suppression controller which is of I-PD structure.

$$U(s) = \frac{1}{A_c(s)}[B_a(s)\omega_M^*(s) - B_c(s)\omega_M(s)], \tag{4.27}$$

where

$$A_c(s) = s, \quad B_c(s) = K_d s^2 + K_p s + K_i, \quad B_a(s) = f_0. \tag{4.28}$$

The two closed-loop transfer functions are given by

$$\frac{\omega_M}{\omega_M^*} = \frac{f_0 K_o (s^2 + \omega_a^2)}{P(s)}, \tag{4.29}$$

$$\frac{\omega_L}{\omega_M^*} = \frac{f_0 K_o \omega_a^2}{P(s)}, \tag{4.30}$$

where $P(s)$ is the characteristic polynomial.

$$P(s) = (1 + K_o K_d)s^4 + K_o K_p^3 + (\omega_r^2 + K_o(\omega_a^2 K_d + K_i))s^2 + K_o K_p \omega_a^2 s + K_o K_i \omega_a^2. \tag{4.31}$$

Based on the CDM design explained in Chap. 3, the characteristic polynomial P can be represented in terms of the stability indexes γ_i and the equivalent time constant τ.

$$P(s) = \left(\frac{\tau^4 a_0}{\gamma_3 \gamma_2^2 \gamma_1^3}\right) s^4 + \left(\frac{\tau^3 a_0}{\gamma_2 \gamma_1^2}\right) s^3 + \left(\frac{\tau^2 a_0}{\gamma_1}\right) s^2 + (\tau a_0)s + a_0. \tag{4.32}$$

From the matching conditions of Eqs. (4.31) and (4.32), it can be seen that the following 5 equations have to be solved for 3 unknown PID gains:

$$1 + K_o K_d = \frac{\tau^4 a_0}{\gamma_3 \gamma_2^2 \gamma_1^3}, \tag{4.33}$$

$$K_o K_p = \frac{\tau^3 a_0}{\gamma_2 \gamma_1^2}, \tag{4.34}$$

$$\omega_r^2 + K_o(\omega_a^2 K_d + K_i) = \frac{\tau^2 a_0}{\gamma_1}, \tag{4.35}$$

$$K_o K_p \omega_a^2 = \tau a_0, \tag{4.36}$$

$$K_o K_i \omega_a^2 = a_0. \tag{4.37}$$

From Eqs. (4.34) and (4.36), we have

$$\tau = \frac{\gamma_1 \sqrt{\gamma_2}}{\omega_a}. \tag{4.38}$$

Substituting Eqs. (4.37) and (4.38) into Eqs. (4.33) and (4.35), respectively, result in

$$K_d - \left(\frac{\gamma_1}{\omega_a^2 \gamma_3}\right) K_i = -\frac{1}{K_o}, \tag{4.39}$$

$$\omega_a^2 K_d - (\gamma_1 \gamma_2 - 1) K_i = -\frac{\omega_r^2}{K_o}. \tag{4.40}$$

Solving Eqs. (4.39) and (4.40) yields

$$K_i = \frac{J_L\, \omega_a^2\, \gamma_3}{(\gamma_1\gamma_2\gamma_3 - \gamma_1 - \gamma_3)}, \tag{4.41}$$

$$K_d = \left[\frac{J_L\gamma_1}{(\gamma_1\gamma_2\gamma_3 - \gamma_1 - \gamma_3)} - \frac{1}{K_o}\right]. \tag{4.42}$$

Then from Eqs. (4.36), (4.37), (4.38), and (4.41), we have

$$K_p = \frac{J_L\omega_a\gamma_1\sqrt{\gamma_2\gamma_3}}{(\gamma_1\gamma_2\gamma_3 - \gamma_1 - \gamma_3)}. \tag{4.43}$$

Finally, if f_0 is obtained so that the closed-loop system becomes Type 1, that is, letting $T(0) = 1$ from Eq. (4.29), it results in

$$f_0 = K_i. \tag{4.44}$$

In this case, since the matching equations of the fourth-order characteristic polynomials have four design parameters (τ and 3 $\gamma_i s$) but three unknowns, τ cannot be arbitrarily selected but is dependent on $\gamma_i s$ as like Eq. (4.38).

Example 4.1 Consider the two-inertia system [8] shown in Fig. 4.7, where the parameters of the system are given by

$$J_M = 4.016 \times 10^{-3}(\text{kg m}^2), \quad J_L = 2.921 \times 10^{-3}(\text{kg m}^2), \quad K_s = 39.21(\text{Nm/rad}).$$

The resonant and anti-resonant frequencies calculated by Eq. (4.26) are $\omega_r = 152.3\,(\text{rad/s})$ and $\omega_a = 115.9\,(\text{rad/s})$, respectively. As a set of the stability index obtaining well damped transient response characteristics, we chose the CDM standard form, which is

$$\gamma_1 = 2.5, \quad \gamma_2 = \gamma_3 = 2.0.$$

Using these stability indexes from Eqs. (4.41)–(4.44), the following I-PD controller gains are determined.

$$[K_p \;\; K_i \;\; K_d \;\; f_0] = [0.4351 \;\; 14.2582 \;\; -0.002688 \;\; 14.2582]. \tag{4.45}$$

From Eq. (4.38), the corresponding equivalent time constant is $\tau = 2.5\sqrt{2}/115.9 = 0.0305$. Step responses of the closed-loop systems Eqs. (4.29) and (4.30) are shown in Fig. 4.9. It is seen that both responses ω_M and ω_L have almost no overshoot and settling time of 0.07 s. Therefore, it is shown that the designed controller successfully suppresses the vibration of the two-inertia system.

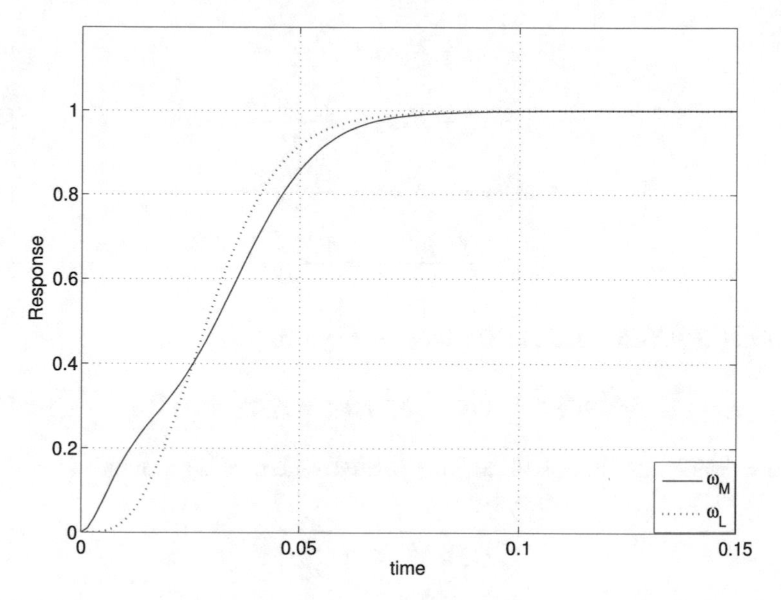

Fig. 4.9 Step responses of the closed-loop systems Eqs. (4.29) and (4.30) with I-PD controller

4.2.2 Design of a First-Order Controller (FOC)

We now design a first-order controller using CDM for the vibration suppression control system shown in Fig. 4.8. The control law in two parameter configuration is

$$U(s) = \frac{1}{A_c(s)}[B_a(s)\omega_M^*(s) - B_c(s)\omega_M(s)], \tag{4.46}$$

where

$$A_c(s) = s + x_1, \quad B_c(s) = x_2\, s + x_3, \quad B_a(s) = f_0. \tag{4.47}$$

The closed-loop transfer functions from ω_M^* to ω_M and ω_L are the same as those in Eqs. (4.29) and (4.30) except the characteristic polynomial, which is

$$P(s) = s^4 + (x_1 + K_o x_2)s^3 + (\omega_r^2 + K_o x_3)s^2 + (\omega_r^2 x_1 + K_o \omega_a^2 x_2)s + K_o \omega_a^2 x_3. \tag{4.48}$$

Rewriting the characteristic polynomial Eq. (4.32) in a monic polynomial form, we have

$$P(s) = s^4 + \left(\frac{\gamma_1\gamma_2\gamma_3}{\tau}\right)s^3 + \left(\frac{\gamma_1^2\gamma_2^2\gamma_3}{\tau^2}\right)s^2 + \left(\frac{\gamma_1^3\gamma_2^2\gamma_3}{\tau^3}\right)s + \left(\frac{\gamma_1^3\gamma_2^2\gamma_3}{\tau^4}\right). \tag{4.49}$$

The CDM approach is to match Eqs. (4.48) to (4.49). It results in the following four equalities for 3 unknown FOC gains $\{x_1, x_2, x_3\}$.

$$x_1 + K_o x_2 = \frac{\gamma_1\, \gamma_2\, \gamma_3}{\tau}, \tag{4.50}$$

$$\omega_r^2 + K_o x_3 = \frac{\gamma_1^2\, \gamma_2^2\, \gamma_3}{\tau^2}, \tag{4.51}$$

$$x_1 \omega_r^2 + K_o\, \omega_a^2\, x_2 = \frac{\gamma_1^3\, \gamma_2^2\, \gamma_3}{\tau^3}, \tag{4.52}$$

$$\frac{\gamma_1^3\, \gamma_2^2\, \gamma_3}{\tau^4} = \frac{\gamma_1^3\, \gamma_2^2\, \gamma_3}{\tau^4}. \tag{4.53}$$

Equalizing Eqs. (4.51) and (4.53) with $z := \tau^2$ yields

$$(\omega_a^2\, \omega_r^2)\, z^2 - (\omega_r^2\, \gamma_1^2\, \gamma_2^2\, \gamma_3)\, z + \gamma_1^3\, \gamma_2^2\, \gamma_3 = 0. \tag{4.54}$$

The discriminant for Eq. (4.54) to have positive real roots is given by

$$\gamma_1\, \gamma_2^2\, \gamma_3 \geq 4\, \frac{\omega_r^2}{\omega_a^2}. \tag{4.55}$$

Now, suppose that stability indexes are selected so that the discriminant is satisfied. Let the larger one of two roots of Eq. (4.54) be τ_h and τ_l for the smaller one. Choose either τ_h or τ_l as the equivalent time constant, i.e., $\tau_x = \tau_h$ (or τ_l). As explained in Chap. 3, the smaller τ_x, the faster the time response.

Then solving Eqs. (4.50) and (4.52) results in

$$x_1 = \frac{1}{(\omega_r^2 - \omega_a^2)} \left(\frac{\gamma_1^2\, \gamma_2^2\, \gamma_3}{\tau_x^3} - \frac{\omega_a^2\, \gamma_1\, \gamma_2\, \gamma_3}{\tau_x} \right), \tag{4.56}$$

$$x_2 = \frac{1}{K_o} \left(\frac{\gamma_1\, \gamma_2\, \gamma_3}{\tau_x} - x_1 \right). \tag{4.57}$$

And from Eqs. (4.53) or (4.51), we have

$$x_3 = \frac{1}{K_o \omega_a^2}\, \frac{\gamma_1^3\, \gamma_2^2\, \gamma_3}{\tau_x^4}, \quad \text{or}$$
$$= \frac{1}{K_o} \left(\frac{\gamma_1^2\, \gamma_2^2\, \gamma_3}{\tau_x^2} - \omega_r^2 \right). \tag{4.58}$$

If f_o is obtained so that the closed-loop system becomes Type 1, from Eqs. (4.29) and (4.48), we have

$$F(s) = f_o = x_3. \tag{4.59}$$

As in the case of I-PD controller in the previous subsection, it should be noted that in the case of FOC, τ_x cannot be specified as an arbitrary value due to the lower-order controller. If we want to specify both desired damping and response speed, it can be solved by increasing the order of controller by, for example, a second-order.

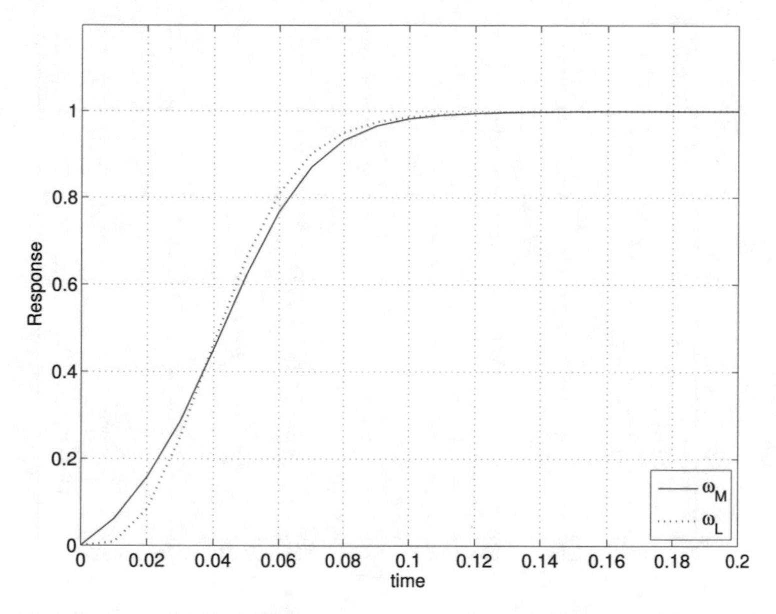

Fig. 4.10 Step responses of the closed-loop systems Eqs. (4.29) and (4.30) with FOC Eq. (4.60)

Example 4.2 Consider the same two-inertia system as the model in Example 4.1. Suppose that the design purpose is to find the FOC that satisfies the no overshoot performance. As a candidate set of stability index that satisfies no overshoot performance, let's first choose the CDM standard form as follows:

$$\gamma_1 = 2.5, \quad \gamma_2 = \gamma_3 = 2.0.$$

The above design parameters satisfy the discriminant Eq. (4.55). Two roots of Eq. (4.54) to τ are

$$[\tau_h \quad \tau_l] = [0.04416 \quad 0.01435].$$

We first select the larger value of them as the equivalent time constant, i.e., $\tau_x = \tau_h = 0.04416$. Then, from Eqs. (4.56)–(4.59), the following FOC gains are determined.

$$[x_1 \quad x_2 \quad x_3 \quad f_0] = [-162.68 \quad 1.5627 \quad 9.83 \quad 9.83]. \tag{4.60}$$

Step responses of the closed-loop systems from ω_M^* to ω_M and ω_L are shown in Fig. 4.10. It is seen that both responses ω_M and ω_L have no overshoot and settling time of 0.098 s. Comparing this FOC result with I-PD one, while the FOC design with $\tau_x = \tau_h$ produced better transient response than I-PD, the controller itself was unstable.

Second, let's try the FOC design by selecting the smaller τ_x, i.e., $\tau_x = \tau_l = 0.01435$. Similarly, Eqs. (4.56)–(4.59) yield

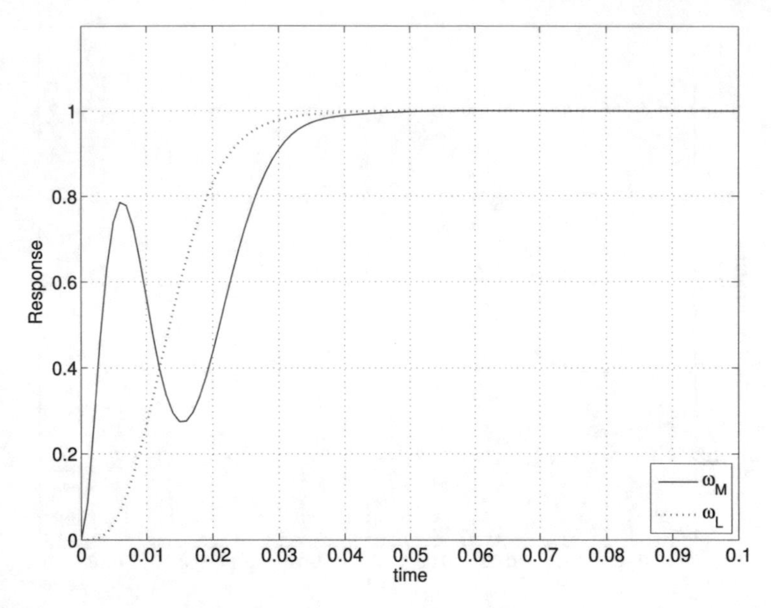

Fig. 4.11 Step responses of the closed-loop systems Eqs. (4.29) and (4.30) with FOC Eq. (4.61)

$$[x_1 \quad x_2 \quad x_3 \quad f_0] = [3375.2 \quad -10.756 \quad 882.11 \quad 882.11]. \tag{4.61}$$

In this case, the resulting FOC is stable. Figure 4.11 shows the step responses of the corresponding closed-loop systems. Both responses have almost no overshoot (0.01%) and settling time of 0.0366 s. Since τ_x is about 1/3 times smaller than in the first case, the responses are 3 times faster, but the transient response of ω_M is much worse.

4.3 A Solution to the ACC Benchmark Problem Using CDM

A benchmark problem for robust control design was proposed by Wie and Bernstein [9] in American Control Conference (ACC) and has been examined in numerous journal articles [10]. The plant of interest is a two-mass-spring system consisting of two mobile carts interconnected with a spring. This section outlines the application of CDM to solve the ACC benchmark problem.

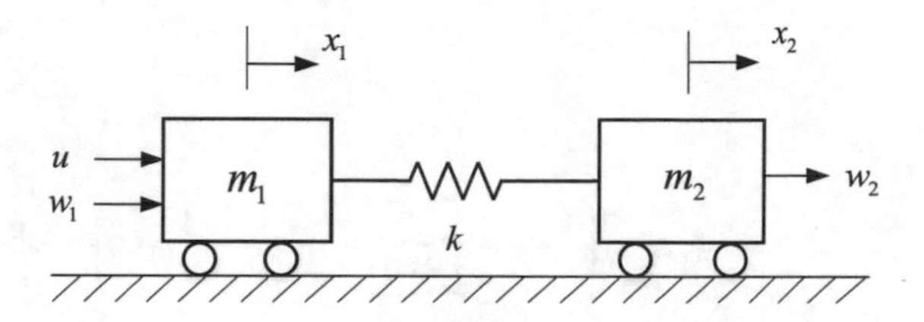

Fig. 4.12 Two-mass-spring system with uncertain parameters

4.3.1 Benchmark Problems

Consider the two-mass-spring system shown in Fig. 4.12, which is a generic model of an uncertain dynamical system with a rigid body mode and on vibration mode. It is assumed that for the nominal system $m_1 = m_2 = 1$ and $k = 1$ with appropriate units. A control force acts on body 1, and the position of body 2 is measured, resulting in a noncollocated actuator/sensor control problem.

The system can be represented in state-space form as

$$\begin{bmatrix} \dot{x}_1 \\ \dot{x}_2 \\ \dot{x}_3 \\ \dot{x}_4 \end{bmatrix} = \begin{bmatrix} 0 & 0 & 1 & 0 \\ 0 & 0 & 0 & 1 \\ -k/m_1 & k/m_1 & 0 & 0 \\ k/m_2 & -k/m_2 & 0 & 0 \end{bmatrix} \begin{bmatrix} x_1 \\ x_2 \\ x_3 \\ x_4 \end{bmatrix} + \begin{bmatrix} 0 \\ 0 \\ 1/m_1 \\ 0 \end{bmatrix} (u + w_1) + \begin{bmatrix} 0 \\ 0 \\ 0 \\ 1/m_2 \end{bmatrix} w_2 \tag{4.62}$$

$$y = x_2 + v \tag{4.63}$$

$$z = x_2 \tag{4.64}$$

where x_1 and x_2 are the positions of body 1 and body 2, respectively; x_3 and x_4 are the velocities of body 1 and body 2, respectively; u the control input acting on body 1; y the sensor output; w_1 and w_2 the plant disturbances acting on body 1 and body 2, respectively; v the sensor noise; and z the output to be controlled.

There are four separate problems in the original benchmark problem. However, these problems can be collectively expressed by combining Problems 1 and 2 (see [10]) as follows:

(i) For a unit impulse disturbance exerted on body 1 and body 2, the control output (x_2) of the nominal system shall not exceed 0.1 after 15 s.
(ii) For the same disturbances the peak control level ($|u_{max}|$) of the nominal system shall not exceed 1.
(iii) The gain margin shall be 6 dB or greater and the phase margin shall be at least 30o.
(iv) The closed-loop system shall be stable for $0.5 \leq k \leq 2.0$ when $m_l = m_2 = 1$.
(v) The closed-loop system shall be stable for simultaneous changes $1 - pm \leq k, m_1, m_2 \leq 1 + pm, \quad pm = 0.3$.
(vi) There shall be reasonable high frequency sensor noise rejection, performance robustness, and controller complexity.

By Laplace transform of Eq. (4.62), the dynamic equation for x_2 can be derived as

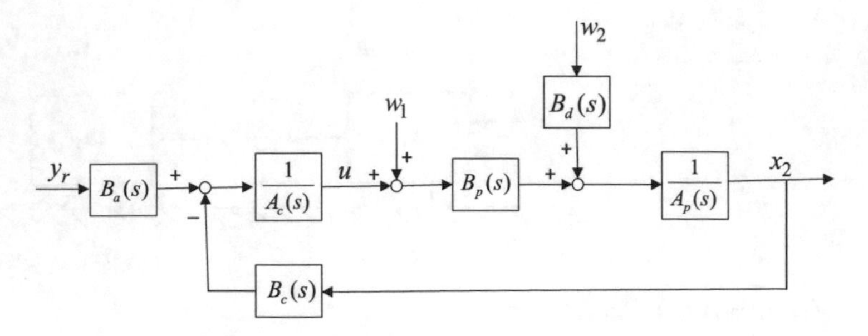

Fig. 4.13 Controller configuration for ACC two-mass-spring system

$$A_p(s)\,x_2 = B_p(s)(u + w_1) + B_d(s)w_2, \tag{4.65}$$

where

$$\begin{aligned} A_p(s) &= (m_1 m_2/k)s^4 + (m_1 + m_2)s^2 = \alpha s^4 + \beta s^2 \\ B_p(s) &= 1 \\ B_d(s) &= (m_1/k)s^2 + 1. \end{aligned}$$

In CDM, the controller structure usually uses the following two parameter configuration:

$$A_c(s)\,u = B_a(s)y_r - B_c(s)(x_2 + v), \tag{4.66}$$

where y_r is the reference input for x_2.

The closed-loop system is given by

$$\begin{bmatrix} x_2 \\ u \end{bmatrix} = \frac{1}{P(s)}\begin{bmatrix} B_p(s) \\ A_p(s) \end{bmatrix}[B_a(s)y_r - B_c(s)v] + \frac{1}{P(s)}\begin{bmatrix} A_c(s) \\ -B_c(s) \end{bmatrix}[B_p(s)w_1 + B_d(s)w_2], \tag{4.67}$$

where $P(s)$ is the characteristic polynomial given by

$$P(s) = A_c(s)A_p(s) + B_c(s)B_p(s). \tag{4.68}$$

The block diagram of the overall control system is shown in Fig. 4.13.

4.3.2 Design Process of CDM Controllers

It is well-known that the Diophantine equation in Eq. (4.68) has a unique solution only if the order of controller satisfies $deg(B_c) = deg(A_p) - 1$ and $deg(A_c) \geq deg(B_c)$. We here consider the third-order controller of the following form:

$$A_c(s) = l_3 s^3 + l_2 s^2 + l_1 s + l_0 \tag{4.69}$$
$$B_c(s) = k_3 s^3 + k_2 s^2 + k_1 s + k_0 \tag{4.70}$$
$$B_a(s) = k_0. \tag{4.71}$$

Without loss of generality, l_0 is assumed to be fixed as $l_0 = 1$. Then the characteristic polynomial becomes

$$\begin{aligned} P(s) &= a_7 s^7 + a_6 s^6 + \cdots + a_1 s + a_0 \\ &= \alpha l_3 s^7 + \alpha l_2 s^6 + (\alpha l_1 + \beta l_3) s^5 + (\alpha + \beta l_2) s^4 \\ &\quad + (\beta l_1 + k_3) s^3 + (\beta + k_2) s^2 + k_1 s + k_0, \end{aligned} \tag{4.72}$$

where $\alpha = m_1 m_2 / k$ and $\beta = m_1 + m_2$.

Remember that in CDM approach, the parameters τ and γ_i are properly selected to achieve the design objectives. As explained in Chap. 2, the reference characteristic polynomial $P(s)$ can be composed of the terms of γ_i and τ. When $n = 7$, for an arbitrary a_0,

$$a_1 = \tau a_0, \quad a_i = \tau^i a_0 / (\gamma_{i-1} \gamma_{i-2}^2 \cdots \gamma_2^{i-2} \gamma_1^{i-1}), \quad \text{for } i = 2, 3, \cdots, 6. \tag{4.73}$$

Thus, if $P(s)$ of degree 7 is generated by choosing 7 parameters $(\gamma_1, \cdots, \gamma_6, \tau)$, the CDM design is equivalent to the problem of solving the Diophantine equation Eq. (4.68) for 7 unknown parameters of the controller.

From the problem item (i), the equivalent time constant $\tau = 6$ s is initially selected because the settling time is about 2.5 τ and the required settling time is 15 s. For the nominal case, $\alpha,\ \beta \in [1\ \ 2]$, and for the uncertain $k \in [0.5\ \ 2.0]$, $\alpha \in [0.5\ \ 2.0]$, $\beta = 2$. For the simultaneous changes $0.7 \leq k,\ m_1,\ m_2 \leq 1.3$, the interval bounds of these parameters are $\alpha \in [0.37692\ \ 2.4143]$ and $\beta \in [1.4\ \ 2.6]$ respectively. The gain margin specification of 6 dB can be interpreted as the $\alpha,\ \beta \in [0.5\ \ 1]$.

In some cases, the value of k_2 may be a negative sign. Then it is seen from Eq. (4.72) that the decrease of β will produce much decreased value of a_2, which degrades the stability of the system. For this reason, the design of the k_2 requires special attention. From Eq. (4.67), it can be seen that w_2 requires a much larger control effort compared with w_1, because of the s^2 term. Considering that w_2 is a unit impulse, the peak value of the u will be about k_3/a_6 or k_3/l_2 for the nominal case. Thus, this value is taken as a measure of the control effort. In this CDM design, we are going to have the value k_3/l_2 around 1.

For simplicity, the following shorthand notations are used in this section.

$$\gamma_i = [\gamma_6\ \cdots\ \gamma_2\ \gamma_1],\ a_i = [a_7\ \cdots\ a_1\ a_0],\ k_i = [k_3\ k_2\ k_1\ k_0],\ l_i = [l_3\ l_2\ l_1\ l_0].$$

In order to evaluate the performance of the controllers to be designed, several performance indices are introduced. The first one is the figure of merits (FM) defined as

$$FM = [PM \quad GM \quad t_s \quad u_{max} \quad k_{min}/k_{max} \quad pm], \tag{4.74}$$

where PM is the phase margin, GM the gain margin, t_s the settling time, u_{max} the maximum value of the control effort, k_{min}/k_{max} the variation range of k for which the system remains stable, and pm the range of simultaneous variation on (k, m_1, m_2) for which the system remains stable.

In addition, we define the key figure (KF) as

$$KF = [k_3/l_2 \quad k_2]. \tag{4.75}$$

The first entry of the KF is a measure of the control effort and the second entry is a measure of robustness.

Also, Thompson's scoring index [10] is used as the metric of performance

$$\begin{aligned} score = lim\left[\frac{PM-30}{5}\right] + lim\left[\frac{GM-6}{2}\right] + lim\left[\frac{15-t_s}{3}\right] + lim\left[\frac{pm-0.3}{0.05}\right] \\ +upperlim\left[\frac{-20log_{10}u_{max}}{3}\right] + lim\left[\frac{20log_{10}(k_{max}/k_{min})-12}{3}\right] + bonus, \end{aligned} \tag{4.76}$$

where $lim(x)$ is a function bounding x to within ± 2, $upperlim(x)$ is a function bounding x to a maximum value of $+2$, and a *bonus* score of 2 is granted if all the requirements are satisfied.

(i) Preliminary Design Based on the CDM Standard Form

The CDM initial design, CDM-1, is carried out using the CDM standard form for parameters γ_i and τ, which are

$$\gamma_i = [2 \ 2 \ 2 \ 2 \ 2 \ 2.5], \quad \tau = 6. \tag{4.77}$$

Using the CDM Matlab script `g2c` in Appendix with the above parameters, the CDM controller can be obtained as follows:

```
>> ap=[1 0 2 0 0]; bp=[1]; nc=3; mc=3; t=6.0; unc=0;
>> gr=[2 2 2 2 2 2.5];
>> [bc,ac,aa,g,tau,gs,rr]=g2c(ap,bp,nc,mc,gr,t,unc)
```

Then the CDM-1 controller and its performance indices are given by

$$\begin{aligned} k_i &= [1.187 \quad -0.4738 \quad 0.6359 \quad 0.1060] \\ l_i &= [0.00379 \quad 0.04945 \quad 0.3223 \quad 1.0] \\ a_i &= [0.0037 \ 0.0495 \ 0.3297 \ 1.0989 \ 1.8315 \ 1.5263 \ 0.6359 \ 0.1060] \\ FM &= [39.5 \quad 6.6 \quad 12.9 \quad \mathbf{14.7} \quad 0.35/3.25 \quad 0.39] \\ KF &= [1.187/0.04945 \quad -0.4738] \\ score &= -1.2 \end{aligned}$$

The bold letters in FM indicate performances that do not meet the given specifications. It is seen that this initial design is robust and satisfies all the specifications except the control effort u_{max}, which is fairly large. This shows that decreasing u_{max} and increasing robustness are in a trade-off. In other words, one has to be sacrificed for the other.

(ii) Improvement of the Control Effort

In order to reduce u_{max}, it is necessary to make l_2 large and k_3 small. This can be achieved by reducing the stability index γ_i, especially of the low order. However, decreasing γ_1 and γ_2 should be avoided as it has a dominant effect on the time response. Thus, we decide to reduce γ_3, γ_4, and γ_5 to 1.5. By Lipatov's sufficient condition for stability [11], the closed-loop system is stable if all the γ_i's are larger than 1.5. We choose $\gamma_6 = 2.0$ because it does not affect the value of l_2.

Thus, CDM-2 design is made with

$$\gamma_i = [2\ \ 1.5\ \ 1.5\ \ 1.5\ \ 2\ \ 2.5], \quad \tau = 6. \tag{4.78}$$

Using the CDM Matlab script `g2c` as like the previous CDM-1, the CDM-2 controller and its performance indices are determined as follows:

$$\begin{aligned}
k_i &= [0.5126 \quad -0.3219 \quad 0.6992 \quad 0.1165] \\
l_i &= [0.0543 \quad 0.3055 \quad 0.7506 \quad 1.0] \\
a_i &= [0.0543 \quad 0.3055 \quad 0.8592 \quad 1.6110 \quad 2.0137 \quad 1.6781 \quad 0.6992 \quad 0.1165] \\
FM &= [38.2 \quad 6.2 \quad 12.8 \quad 0.97 \quad \mathbf{0.53/1.74} \quad 0.23] \\
KF &= [0.5126/0.3055 \quad -0.3219] \\
score &= 0.6518
\end{aligned}$$

The CDM-2 controller satisfies the input condition $|u_{max}| \leq 1$, but k_{min}/k_{max} is not satisfied. To find out the cause of instability, let us compare γ_i and γ_i^* for the variation of the spring constant k.

For the lower limit $k = 0.5$, we have

$$\gamma_i = [2.135 \quad 1.624 \quad 2.103 \quad \mathbf{0.9255} \quad 2.0 \quad 2.5] \tag{4.79}$$

$$\gamma_i^* = [0.6156 \quad 0.9539 \quad 1.696 \quad \mathbf{0.9755} \quad 1.481 \quad 0.5] \tag{4.80}$$

It is shown clearly that γ_3 is deteriorated. Since $\gamma_3^* = 1/\gamma_4 + 1/\gamma_2$, γ_3^* can be reduced by increasing γ_2.

Also, for the upper limit $k = 2$

$$\gamma_i = [1.776 \quad \mathbf{1.380} \quad 1.267 \quad 2.175 \quad 2.0 \quad 2.5] \tag{4.81}$$

$$\gamma_i^* = [0.7247 \quad \mathbf{1.353} \quad 1.184 \quad 1.290 \quad 0.8598 \quad 0.5] \tag{4.82}$$

Clearly, γ_5 is deteriorated. γ_5^* can be reduced by increasing γ_6.

(iii) Improvement of the robustness

Based on the above analysis, let us now try to design the CDM-3 controller with increasing γ_2 and γ_6, which are

$$\gamma_i = [4\ \ 1.5\ \ 1.5\ \ 1.5\ \ 2.5\ \ 2.0], \quad \tau = 6. \tag{4.83}$$

Similarly, the CDM Matlab script `g2c` results in the CDM-3 controller, and its performance indices are determined as follows:

$$\begin{aligned}
k_i &= [0.4040\quad -0.3219\quad 0.5594\quad 0.0932]\\
l_i &= [0.0272\quad 0.3055\quad 0.8049\quad 1.0]\\
a_i &= [0.027\ 0.3055\ 0.859\ 1.611\ 2.0137\ 1.678\ 0.559\ 0.093]\\
FM &= [37.4\quad 7.3\quad 14.1\quad 0.89\quad 0.46/2.29\quad 0.35]\\
KF &= [0.404/0.3055\quad -0.3219]\\
score &= 6.45
\end{aligned}$$

It is seen that the CDM-3 controller satisfies all the specifications with comfortable margin. The choice of $\gamma_6 = 4$ is more or less arbitrary. However, it can be argued that increasing γ_6 improves the robustness. As will be shown later, k_3, k_2, and l_2 are function of τ_2, which is

$$\tau_2 = \frac{a_3}{a_2} = \frac{\tau}{\gamma_2\gamma_1} \tag{4.84}$$

For the same τ, the value of τ_2 when $\gamma_1 = 2.5$ and $\gamma_1 = 2.0$ is the same as that of $\gamma_1 = 2.0$ and $\gamma_2 = 2.5$. Thus both cases have the same values of k_2 and l_2. In an ordinary system, the time response without overshoot is preferred. For this purpose, it is recommended to choose $\gamma_1 = 2.5$. In this benchmark problem, the overshoot is not limited but the settling time of 15 s should be satisfied as in the term (i). In accordance with these considerations, we choose $\gamma_2 = 2.5$ as the maximum value and accordingly γ_1 is chosen to be 2 for a good time response.

(iv) Final design

Although the CDM-3 controller is satisfactory, further improvement can be achieved by increasing τ because the settling time has some margin. By increasing τ, the CDM-4 controller is obtained with

$$\gamma_i = [4\ \ 1.5\ \ 1.5\ \ 1.5\ \ 2.5\ \ 2.0], \quad \tau = 6.4 \tag{4.85}$$

Similarly, the CDM Matlab script `g2c` results in the CDM-4 controller, and its performance indices are determined as follows:

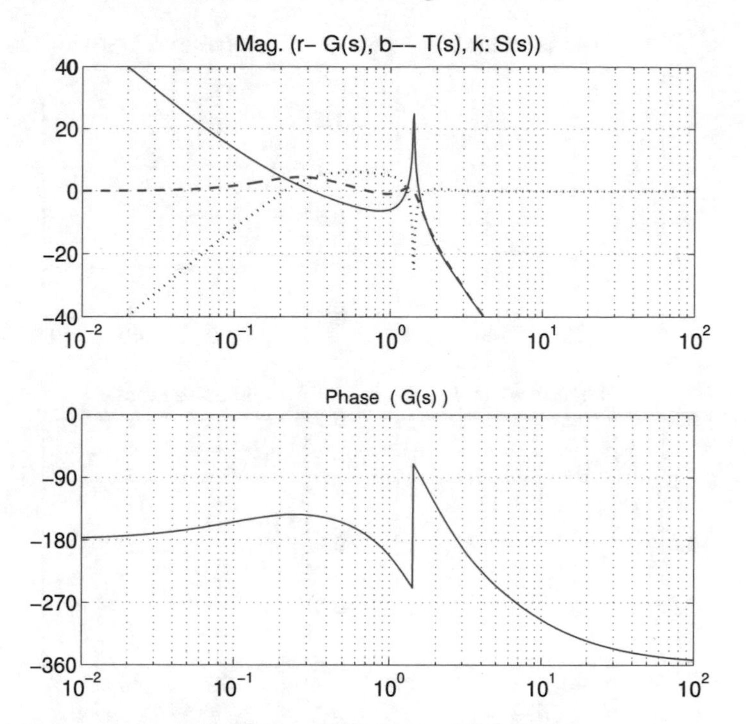

Fig. 4.14 Frequency responses of $G(s) = B_c B_p / A_c A_p$, S(s), and T(s)

$$
\begin{aligned}
k_i &= [0.2039 \quad -0.3895 \quad 0.5033 \quad 0.0786] \\
l_i &= [0.0360 \quad 0.3795 \quad 0.9287 \quad 1.0] \\
a_i &= [0.0360 \; 0.3795 \; 1.0007 \; 1.7591 \; 2.0614 \; 1.6105 \; 0.5033 \; 0.0786] \\
FM &= [35.4 \quad 6.2 \quad 15.0 \quad 0.59 \quad 0.47/3.45 \quad 0.42] \\
KF &= [0.2039/0.3795 \quad -0.3895] \\
score &= 8.54
\end{aligned}
$$

It is seen that the CDM-4 controller satisfies all the specifications with a high score of 8.54. Frequency responses of the loop transfer function, sensitivity and complementary sensitivity functions are shown in Fig. 4.14 and time responses of the closed-loop system from impulse inputs w_1 and w_2 to x_2 and u are shown in Fig. 4.15.

4.3.3 Analysis and Discussion

As seen in the previous section, robustness and control effort are closely related to the CDM design parameters such as stability index and equivalent time constant. The sufficient condition for stability by Lipatov and Sokolov [11] makes it possible

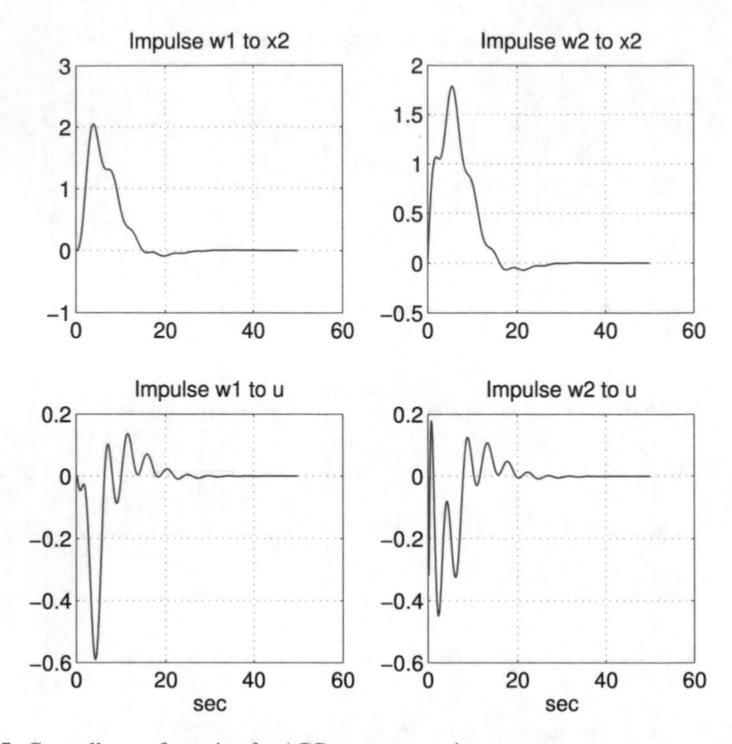

Fig. 4.15 Controller configuration for ACC two-mass-spring system

to design a stable controller by just selecting γ_i's. These characteristics are the most conspicuous points in the CDM. In this section, we will explain how the CDM design parameters have been selected to meet the given performance of the ACC benchmark problem.

First, we will discuss how to select γ_i for gain margin and the maximum input constraint. From coefficient of the characteristic polynomial in Eq. (4.73), we have

$$a_i = a_2\ \tau_2^{i-2}/(\gamma_{i-1}\ \gamma_{i-2}^2\ \cdots\ \gamma_3^{i-3}),\quad i = 3, 4, \cdots, n \tag{4.86}$$

$$\tau_2 = \tau/(\gamma_1\gamma_2),\quad a_2 = a_0\ \tau^2/\gamma_1 \tag{4.87}$$

For the nominal case, from Eqs. (4.72) and (4.73), the controller parameters cane be expressed in terms of γ_i and τ_2 as follows:

$$l_3 = (2 + k_2)\ \tau_2^5/(\gamma_6\ \gamma_5^2\ \gamma_4^3\ \gamma_3^4) \tag{4.88}$$

$$l_2 = (2 + k_2)\ \tau_2^4/(\gamma_5\ \gamma_4^2\ \gamma_3^3) \tag{4.89}$$

$$l_1 + 2l_3 = (2 + k_2)\ \tau_2^3/(\gamma_4\ \gamma_3^2) \tag{4.90}$$

$$1 + l_2 = (2 + k_2)\ \tau_2^2/\gamma_3 \tag{4.91}$$

$$k_3 + 2l_1 = (2 + k_2)\ \tau_2. \tag{4.92}$$

Eliminating l_2 by combining Eqs. (4.89) and (4.91) yields

$$1 = (2 + k_2)[\tau_2^2/\gamma_3 - 2\,\tau_2^4/(\gamma_5\,\gamma_4^2\,\gamma_3^3)]. \tag{4.93}$$

From this equation, it is clear that for the given τ_2, k_2 is dependent upon γ_5, γ_4, and γ_3. For $\tau = 6$ and $\gamma_1\gamma_2 = 5$, then τ_2 is equal to 1.2. So, if $k_2 \geq -0.3$ is assumed for the condition of $G \geq 6\,\text{dB}$, as will be explained later, and if γ_5, γ_4, and γ_3 are chosen to be equal, then the solution of Eq. (4.93) gives

$$\gamma_5 = \gamma_4 = \gamma_3 \leq 1.4909, \quad \text{or} \quad \geq 2.3495. \tag{4.94}$$

This is the limitation imposed on γ_i by GM requirement.

Next, let us discuss about the condition $u_{max} \leq 1$ for a unit impulse disturbance. In order to satisfy this condition, we can see from Eq. (4.67) that the value of k_3/l_2 should be less than or equal to 1. Eliminating l_1 from Eqs. (4.92) and (4.90), and dividing it by Eq. (4.89) gives the following relation:

$$k_3/l_2 = [1 - 2\,\tau_2^2/(\gamma_4\,\gamma_3^2)/[\tau_2^3/(\gamma_5\,\gamma_4^2\,\gamma_3^3)] + 4\,l_3/l_2. \tag{4.95}$$

Assuming that $l_3/l_2 \approx 0$ (this value is usually very small), $\tau_2 = 1.2$, $\gamma_5 = \gamma_4 = \gamma_3$, and k_3/l_2, then the following result is derived.

$$\gamma_5 = \gamma_4 = \gamma_3 \leq 1.5022. \tag{4.96}$$

From Eqs. (4.94) and (4.96), we can conclude that γ_5, γ_4, and γ_3 must take the values of around 1.5 in order to satisfy the conditions for GM and u_{max}. Also, it should be noted that the value $\gamma_5 = \gamma_4 = \gamma_3 = 1.5$ satisfies the sufficient condition for stability, which is $\gamma_i > 1.12\gamma_i^*$. We can see that this benchmark problem is a very tight problem since the solution lies in a very narrow limited region. For example, if $\gamma_5 = \gamma_4 = \gamma_3 \leq 1.4$ is selected, no solution exists for this problem.

Many solutions using various design methods to this benchmark problem have been published. In [12], 11 solutions were presented. Thompson [10] modified the specifications of the original benchmark problem more concretely and suggested his own design using classical/H_2 approach. A survey of benchmark problem solutions are also given in [10]. For the purpose of comparison, some solutions with a high score are shown in Table 4.1. In Thompson's score, the perfect score is 14, and if the score is +4 of higher, it is very good, if it is higher than 0, it is good, and if it is lower than −4, it is in need of improvement.

Table 4.1 Comparison of some benchmark problem solutions

Method	PM	GM	t_s	u_{max}	k_{min}/k_{max}	pm	All	Score
Requirements	30	6.0	15	1	0.5/2.0	0.3		
CDM-4 [13, 14]	35.4	6.2	15.0	0.59	0.47/3.45	0.42	√	8.54
Thompson [10], Improved Wie's H_∞	31.7	6.1	14.9	0.56	0.37/2.53	0.42	√	7.66
Thompson [10], Classical& H_2	35.3	6.0	14.5	0.76	0.45/2.8	0.42	√	7.36
Wie [12], H_∞	34.2	6.1	**15.2**	0.57	0.44/3.91	0.459	×	6.43
Chiang [12], H_∞ & pole shifting	30.2	**3.9**	**15.1**	0.8	0.46/3.70	0.36	×	2.8
Lilja [12], Approx. pole placement	**23.8**	**3.7**	**29.0**	0.55	0.23/∞	0.35	×	0.26
Byrns [12], H_∞ & LTR	**23.3**	**3.0**	14.2	0.88	**0.51**/3.56	0.3	×	−0.62
Braatz [12], μ synthesis	**27.2**	**2.8**	14.1	0.95	**0.57**/2.5	**0.28**	×	−1.9
Jayasuriya [12], QFT	**11**	**2.5**	5.0	**360**	**0.18/1.3**	**0.09**	×	−19.1

All: All requirements are satisfied by (√) and not by (×).
score: Thompson's score by Eq. (4.76)
Bold numbers: indicate that the performances did not meet the given requirements

4.4 A Longitudinal Control of a Modern Aircraft

It has been shown that CDM can be effectively applied mainly to SISO or SIMO control design problems. In this section, an example of the application of CDM to the multi-input multi-output(MIMO) problem is shown. The problem to be considered is taken from the well-known example, which is the longitudinal control of the modern fighter with dual control surfaces [15, 16]. For MIMO control systems, various modern control design techniques have been employed.

At present, LQG and H_2/H_∞ approaches are the most popular control design procedures for MIMO systems. However, these methods have not reached expectations for practical application in the aerospace community due to the following reasons:

(1) Parameter tuning procedures are not provided.
(2) Weight selection rules are not established.
(3) The controller order is unnecessarily high.
(4) Extension to gain scheduling or inclusion of proper saturation of state variable is difficult.

This section will present that such a H_∞ problem for the longitudinal control of a fighter can be solved by CDM without difficulty. The results will reveal that the fifth-order CDM controller has a similar performance to the eighth-order H_∞ controller by Chiang [15].

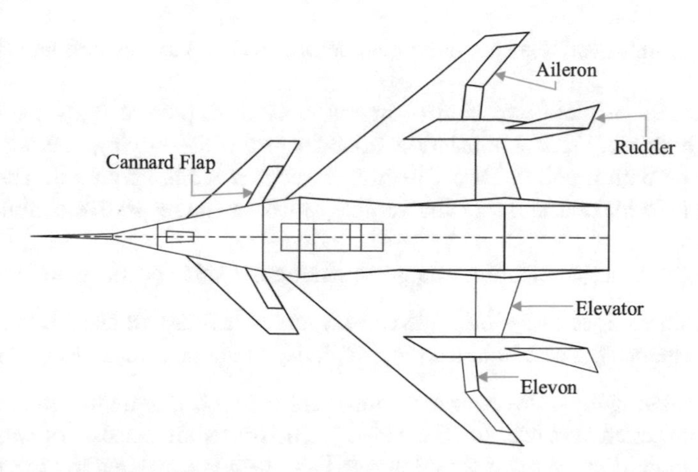

Fig. 4.16 Aircraft configuration

4.4.1 A Fighter Model and Problem Statement

We consider the longitudinal dynamic model of a modern aircraft [15], shown in Fig. 4.16. This aircraft was trimmed at 25000 ft and 0.9 Mach and the linear model in state-space expression is given as

$$\dot{x} = A_g x + B_g u \tag{4.97}$$

$$y = C_g x, \tag{4.98}$$

where $x = [\delta V \ \ \alpha \ \ q \ \ \theta \ \ \delta_e \ \ \delta_c]^T, \quad u = [u_e \ \ u_c]^T, \quad y = [\alpha \ \ \theta]^T,$

$$A_g = \begin{bmatrix} -0.22567 & -36.617 & -18.897 & -32.09 & 3.2509 & -0.76257 \\ 9.2572e-5 & -1.8997 & 0.98312 & -7.2562e-4 & -0.1708 & 0.49652e-3 \\ 0.012338 & 11.72 & -2.6316 & 8.7582e-4 & -31.604 & 22.396 \\ 0 & 0 & 1 & 0 & 0 & 0 \\ 0 & 0 & 0 & 0 & -30 & 0 \\ 0 & 0 & 0 & 0 & 0 & -30 \end{bmatrix}$$

$$B_g^T = \begin{bmatrix} 0 & 0 & 0 & 0 & 30 & 0 \\ 0 & 0 & 0 & 0 & 0 & 30 \end{bmatrix},$$

$$C_g = \begin{bmatrix} 0 & 1 & 0 & 0 & 0 & 0 \\ 0 & 0 & 0 & 1 & 0 & 0 \end{bmatrix}.$$

The state variables are velocity deviation (δV), angle of attack (α) which is the angle between the velocity vector and the x axis of the vehicle, attitude rate (q), attitude angle (θ) which is the angle between the horizontal line and the x axis of the vehicle, elevon angle (δ_e), and canard angle (δ_c). The output variables are α and θ.

The control input variables are elevon actuator input (u_e) and canard actuator input (u_c).

By the use of these two control inputs, very high precise flight path control becomes possible. Vertical translation mode keeps θ while varying α. Pitch pointing mode keeps both α and θ. Direct lift mode keeps α while varying θ. The control objective is to make α and θ of the vehicle to follow the respective commands (α_r and θ_r).

In [15], design specification was given in singular value critria as follows:

(1) Robustness Spec.: 40 dB/decade roll-off and at least 20 dB at 100 rad/s.
(2) Performance Spec.: Minimize the sensitivity function as much as possible.

Before designing the CDM controller, the plant model is represented in the form of the right polynomial matrix fraction (PMF). The procedure consists of two steps: a left PMF model for the plant is first derived and then is converted to the right PMF form. When actuator dynamics are moved to controller, the control inputs becomes δ_e and δ_c. Also q is replaced by $s\theta$, and δV is eliminated from the equations in Eq. (4.97). The left PMF for plant is given as follows:

$$A_p(s)\begin{bmatrix}\alpha\\ \theta\end{bmatrix} = B_u(s)\begin{bmatrix}\delta_e\\ \delta_c\end{bmatrix}, \tag{4.99}$$

where

$$A_p(s) = \begin{bmatrix}a_{p11} & a_{p12}\\ a_{p21} & a_{p22}\end{bmatrix}, \quad B_u(s) = \begin{bmatrix}b_{u11} & b_{u12}\\ b_{u21} & b_{u22}\end{bmatrix},$$
$$a_{p11} = s + 1.9876, \quad a_{p12} = -7.503e-3\,s^2 - 1.0029\,s + 0.73219e-3,$$
$$a_{p21} = -11.72s + 0.1873, \quad a_{p22} = s^3 + 2.6542\,s^2 + 0.29166\,s + 0.39591,$$
$$b_{u11} = 0.066325, \quad b_{u12} = -0.1730$$
$$b_{u21} = -31.604\,s - 0.6731, \quad b_{u22} = 22.396\,s + 0.4960.$$

In order to make design easier, fictitious inputs δ_e^* and δ_c^* are introduced, such that

$$B_u(s)\begin{bmatrix}\delta_e\\ \delta_c\end{bmatrix} = B_p(s)\begin{bmatrix}\delta_c^*\\ \delta_e^*\end{bmatrix}, \tag{4.100}$$

$$B_p(s) = \begin{bmatrix}b_{p11} & 0\\ 0 & b_{p22}\end{bmatrix}, \quad b_{p11} = b_{u12}, \quad b_{p22} = b_{u21}.$$

Then the following relation is derived.

$$\begin{bmatrix}\delta_e\\ \delta_c\end{bmatrix} = E_1\begin{bmatrix}\delta_c^*\\ \delta_e^*\end{bmatrix}, \tag{4.101}$$

where

$$E_1 = B_u^{-1}(s)B_p(s) = \begin{bmatrix} 0.97298 & 1.3730 \\ 1.3730 & 0.52638 \end{bmatrix} + \Delta(s)$$

In this equation the term $\Delta(s)$ can be neglected because it is very small. Thus, the left PMF form of the plant can be expressed by

$$A_p(s)\begin{bmatrix} \alpha \\ \theta \end{bmatrix} = B_p(s)\begin{bmatrix} \delta_c^* \\ \delta_e^* \end{bmatrix}. \tag{4.102}$$

In order to convert Eq. (4.102) to right PMF, new variables α_1 and θ_1 are introduced such that

$$\begin{bmatrix} \alpha \\ \theta \end{bmatrix} = B_p(s)\begin{bmatrix} \alpha_1 \\ \theta_1 \end{bmatrix}. \tag{4.103}$$

Then Eq. (4.102) becomes

$$A_p^*(s)\begin{bmatrix} \alpha_1 \\ \theta_1 \end{bmatrix} = \begin{bmatrix} \delta_c^* \\ \delta_e^* \end{bmatrix}. \tag{4.104}$$

where

$$\begin{aligned} A_p^*(s) &= B_p^{-1}(s)A_p(s)B_p(s) = \begin{bmatrix} a_{p11} & a_{p12}^* \\ a_{p21}^* & a_{p11} \end{bmatrix}, \\ a_{p12}^* &= a_{p12}(b_{p22}/b_{p11}) = -1.370s^3 - 182.23\,s^2 - 3.7681s + 2.8487e - 3, \\ a_{p21}^* &= a_{p21}(b_{p11}/b_{p22}) \simeq -0.064156. \end{aligned}$$

In the end, the right PMF form of the plant Eq. (4.97) is expressed by Eqs. (4.103) and (4.104). Since the outputs of the controller in this form are δ_e^* and δ_c^*, it is necessary to convert them into the actual control inputs u_e and u_c.

$$\begin{bmatrix} u_e \\ u_c \end{bmatrix} = E_1\begin{bmatrix} (s/30+1)\delta_c^* \\ (s/30+1)\delta_e^* \end{bmatrix}. \tag{4.105}$$

4.4.2 *Design of Decoupling Controller*

As mentioned previously, the purpose of control is to make the outputs α and θ to follow the command α_r and θ_r. The specification given in terms of singular value can be interpreted as follows:

(1) Each control channel should be independent and no interaction is expected.
(2) Each channel should have the same characteristics.
(3) The complementary sensitivity function of each channel should have −40 dB/dec roll-off and at least −20 dB at 100 rad/s.

Usually, the sensitivity function becomes larger when the interaction exists between two channels. Thus the minimization of sensitivity function reduces the interaction. The following controller is considered:

$$A_c(s)\begin{bmatrix}\delta_c^*\\ \delta_e^*\end{bmatrix} = B_c(s)\begin{bmatrix}\alpha_r-\alpha\\ \theta_r-\theta\end{bmatrix}, \tag{4.106}$$

where $A_c(s) = A_c^*(s)A_c^{**}$.

$Ac(s)$ consists of two parts. $A_c^*(s)$ is a diagonal channel controller, and $A_c^{**}(s)$ is a decoupling controller, which makes $A_c^{**}(s)A_p(s)$ almost diagonal. Also, we set the reference numerator $B_a(s)$ equal to the feedback numerator $B_c(s)$.

The closed-loop is given as

$$\begin{bmatrix}\alpha\\ \theta\end{bmatrix} = B_p(s)A^{-1}B_c(s)\begin{bmatrix}\alpha_r\\ \theta_r\end{bmatrix}, \tag{4.107}$$

where

$$A(s) = A_c^*(s)A_c^{**}(s)A_p^*(s) + B_c(s)B_p(s). \tag{4.108}$$

$A_c^*(s)$, $B_p(s)$, and $B_c(s)$ are diagonal, and $A_c^*(s)A_p(s)$ is almost diagonal. Because $B_c(s)B_p(s)$ is very large and diagonal, $A(s)$ becomes practically diagonal. Thus, the closed-loop response is decoupled. Decoupling controller can be designed by approximating $adj(A_p^*(s))$ at high order. The following decoupling controller was selected so that anti-diagonal elements of $A_c^{**}(s)A_p^*(s)$ became zero or the higher order terms were eliminated:

$$A_c^{**}(s) = \begin{bmatrix}s+2.6542 & 1.37306s+182.23\\ 0.064156 & s+1.9876\end{bmatrix}. \tag{4.109}$$

Then it is seen that

$$A_{cp} = A_c^{**}(s)A_p^*(s) = \begin{bmatrix}a_{cp11} & a_{cp12}\\ a_{cp21} & a_{cp22}\end{bmatrix}, \tag{4.110}$$

$$a_{cp11} = s^2+4.5539s-6.4157, \quad a_{cp12} = -3.3684s^2+43.693s+72.154$$
$$a_{cp21} = 0, \quad a_{cp22} = s^4+4.5539s^3-6.124s^2+0.73387s+0.78709.$$

Thus, Eq. (4.108) can be written as

$$A(s) = A_c^*(s)A_{cp}(s) + B_c(s)B_p(s), \tag{4.111}$$

where

$$A_c^*(s) = \begin{bmatrix}a_{c11}^* & 0\\ 0 & a_{c22}^*\end{bmatrix}, \quad B_c(s) = \begin{bmatrix}b_{c11} & 0\\ 0 & b_{c22}\end{bmatrix}.$$

Design of controller will proceed on the basis that $A(s)$ is diagonal. This makes it possible to apply the SISO CDM approach to the MIMO problem.

4.4.3 Design of Feedback Controller

From Eqs. (4.110) and (4.111), the open-loop transfer function for α control loop is given by

$$G_\alpha(s) = \frac{b_{c11}\, b_{p11}}{a_{c11}^*\, a_{cp11}}. \tag{4.112}$$

Since the plant is of the second order and integral control needs to be added, the order of controller for α loop should be at least 2. Here the controller is selected as the form

$$a_{c11}^* = l_2 s^2 + s, \quad b_{c11} = k_2 s^2 + k_1 s + k_0, \quad l_2 = 1/30. \tag{4.113}$$

wherein $l_2 = 1/30$ is for actuator dynamics moved to controller transfer function. The closed-loop characteristic polynomial for α loop is given by

$$P_\alpha = s^4 + 34.554s^3 + (130.2 - 5.19k_2)s^2 - (192.47 + 5.19k_1)s - 5.19k_0. \tag{4.114}$$

Design was made on the following choices:

$$\tau_2 = 0.045, \quad \gamma_2 = 4, \quad \gamma_1 = 2.5, \quad \tau = \gamma_2\gamma_1\tau_2 = 0.45. \tag{4.115}$$

The choice of $\gamma_2 = 4$ is to reduce the gain peak at the crossover frequency. The results are as follows:

$$\begin{aligned} &k_2 = -122.88, \quad k_1 = -859.4, \quad k_0 = -1828.6, \quad l_2 = 1/30, \\ &\gamma_i = [1.5549 \ \ 4 \ \ 2.5], \quad \tau = 0.45, \\ &\omega_c \simeq k_2(-0.1730) = 21.257. \end{aligned} \tag{4.116}$$

The ω_c denotes the gain crossover frequency of $G_\alpha(j\omega)$. The selection of $\tau_2 = 0.045$ is to make ω_c around 20 rad/s. The coefficient diagram is shown in Fig. 4.17. The design can be also carried out in constructing the coefficient diagram by hand or by CDM Toolbox in Appendix.

Next, let us consider the open-loop transfer function for θ control loop from Eqs. (4.110) and (4.111).

$$G_\theta(s) = \frac{b_{c22}\, b_{p22}}{a_{c22}^*\, a_{cp22}}. \tag{4.117}$$

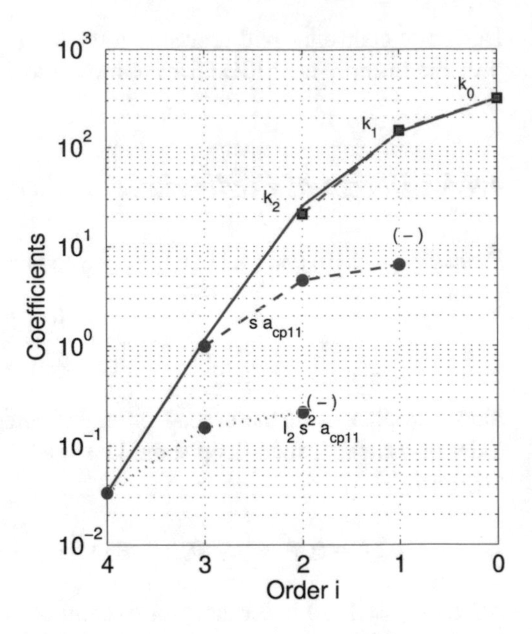

Fig. 4.17 Coefficient diagram for α control

Since the plant is of the fourth order (see a_{cp22} in Eq. (4.110)) and integral control needs to be added, the order of controller for θ loop should be at least 4. However, because the coefficient a_0 of the characteristic polynomial is to be accepted as it is, the order of θ controller is chosen by 3. Here the controller structure was selected as the form

$$a^*_{c22} = (l^*_2 s + 1)(l^*_3 s^2 + s), \quad b_{c22} = k^*_3 s^3 + k^*_2 s^2 + k^*_1 s + k^*_0, \quad l^*_2 = 1/30. \tag{4.118}$$

wherein $l^*_2 = 1/30$ is for actuator dynamics moved to controller transfer function. The closed-loop characteristic polynomial for θ loop is given by

$$\begin{aligned} P_\theta = {} & l^*_2 l^*_3 s^7 + (4.5539 l^*_2 l^*_3 + l^*_2 + l^*_3)s^6 + (1 - 6.124 l^*_2 l^*_3 + 4.5539(l^*_2 + l^*_3))s^5 \\ & +(4.5539 - 0.73387 l^*_2 l^*_3 - 6.124(l^*_2 + l^*_3) - 31.604k_3)s^4 + (0.78709 l^*_2 l^*_3 \\ & +0.73387(l^*_2 + l^*_3) - 6.124 - 31.604k_2 - 0.6731k_3)s^3 + (0.73887 \\ & +0.78709(l^*_2 + l^*_3) - 31.604k_1 - 0.6731k_2)s^2 + (0.78709 - 31.604k_0 \\ & -0.6731k_1)s - 0.6731k_0. \end{aligned} \tag{4.119}$$

Design was made on the following choices:

$$\tau_4 = 0.045, \quad \gamma_4 = 3.2, \quad \gamma_1 = 2.5, \quad \tau_1 = \gamma_4\gamma_3\gamma_2\tau_4 = 0.9. \tag{4.120}$$

With $l^*_3 = 0.003$, the results are as follows:

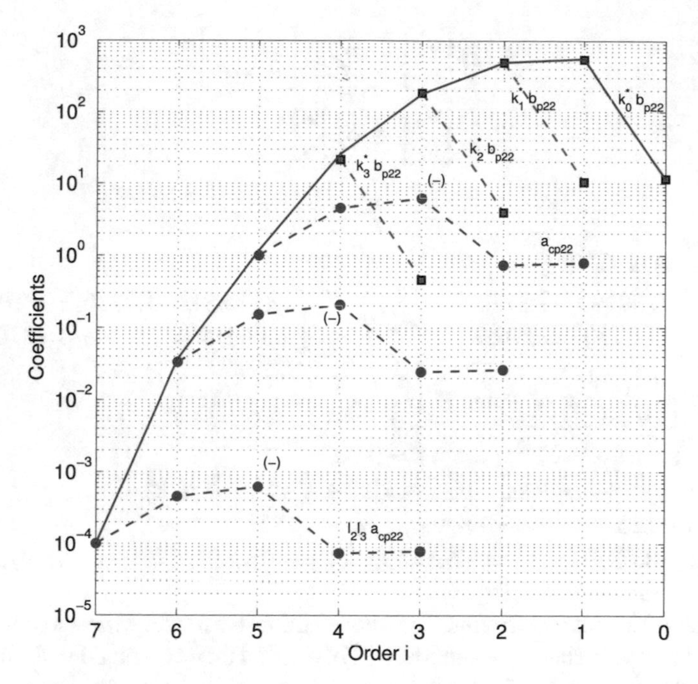

Fig. 4.18 Coefficient diagram for θ control

$$k_3^* = -0.68207, \quad k_2^* = -5.8683, \quad k_1^* = -15.651, \quad k_0^* = -17.197, \tag{4.121}$$

$$l_2^* = 1/30, \quad l_3^* = 0.003, \tag{4.122}$$

$$\gamma_i = [11.619 \quad 1.4248 \quad 3.2 \quad 2.5 \quad 2.5 \quad 53.257],$$

$$\tau_1 = 0.9, \quad \tau = 47.931$$

$$\omega_c \simeq k_3^{*}(-31.604) = 21.556.$$

When plotting the coefficient diagram, l_3 needs to be sufficiently small compared with $l_2^* = 1/30$, and based on this $l_3^* = 0.003$ was chosen heuristically. The coefficient diagram is shown in Fig. 4.18. Due to the zero of the plant near origin, the coefficient of the 0th order can not be designed and has to be accepted as it is.

4.4.4 Results

To evaluate the performance of the designed CDM controller, frequency response and singular value plots will be shown. In this simulation, the original state-space model Eq. (4.97) is used for the plant model. The actuator dynamics in the designed controller are moved to the plant. Then the controller is given as follows:

$$A_c(s)\begin{bmatrix}u_c^*\\u_e^*\end{bmatrix} = B_c(s)\begin{bmatrix}\alpha_r-\alpha\\\theta_r-\theta\end{bmatrix}, \tag{4.123}$$

$$\begin{bmatrix}u_e\\u_c\end{bmatrix} = E_1\begin{bmatrix}u_c^*\\u_e^*\end{bmatrix}, \tag{4.124}$$

where

$$\begin{aligned}
A_c(s) &= A_c^*(s)A_c^{**}(s),\\
E_1 &= \begin{bmatrix}0.97298 & 1.3730\\ 1.3730 & 0.52638\end{bmatrix},\quad A_c^{**}(s) = \begin{bmatrix}s+2.6542 & 1.37306s+182.23\\ 0.064156 & s+1.9876\end{bmatrix},\\
A_c^*(s) &= \begin{bmatrix}a_{c11}^* & 0\\ 0 & a_{c22}^*\end{bmatrix},\quad B_c(s) = \begin{bmatrix}b_{c11} & 0\\ 0 & b_{c22}\end{bmatrix},\\
a_{c11}^* &= s,\quad a_{c22}^* = l_3^* s^2 + s,\quad l_3^* = 0.003,\\
b_{c11} &= k_2 s^2 + k_1 s + k_0,\quad b_{c22} = k_3^* s^3 + k_2^* s^2 + k_1^* s + k_0^*,\\
k_2 &= -122.88,\quad k_1 = -859.4,\quad k_0 = -1828.6,\\
k_3^* &= -0.68207,\quad k_2^* = -5.8683,\quad k_1^* = -15.651,\quad k_0^* = -17.197.
\end{aligned}$$

This is the fifth-order controller. The above controller model expressed in left PFD can be converted to state-space model by Poly-x, and then combined with state-space model of the plant by MATLAB.

The sensitivity function $S(s)$ and complementary sensitivity function $T(s)$ are defined as follows:

$$\begin{bmatrix}\alpha\\\theta\end{bmatrix} = T(s)\begin{bmatrix}\alpha_r\\\theta_r\end{bmatrix}, \tag{4.125}$$

$$\begin{bmatrix}\alpha_r-\alpha\\\theta_r-\theta\end{bmatrix} = S(s)\begin{bmatrix}\alpha_r\\\theta_r\end{bmatrix}, \tag{4.126}$$

where

$$T(s) = \begin{bmatrix}T_{11} & T_{12}\\ T_{21} & T_{22}\end{bmatrix},\quad S(s) = \begin{bmatrix}S_{11} & S_{12}\\ S_{21} & S_{22}\end{bmatrix}.$$

The following relations hold.

$$T_{11} + S_{11} = T_{22} + S_{22} = 1, \tag{4.127}$$

$$T_{12} + S_{12} = T_{21} + S_{21} = 0. \tag{4.128}$$

Frequency responses of the sensitivity and complementary sensitivity functions are shown in Fig. 4.19. It is seen that the cross-coupling terms are very small and invisible on the figures. The singular value plots of both functions are shown in Fig. 4.20. They are almost identical with the frequency response.

The gamma value achieved by the eighth-order H_∞ controller in [15] is 16.8. The corresponding gamma value by the fifth-order CDM controller is about 13 as read from Fig. 4.20. But if the peak around crossover frequency is neglected, the gamma value go up to 16.

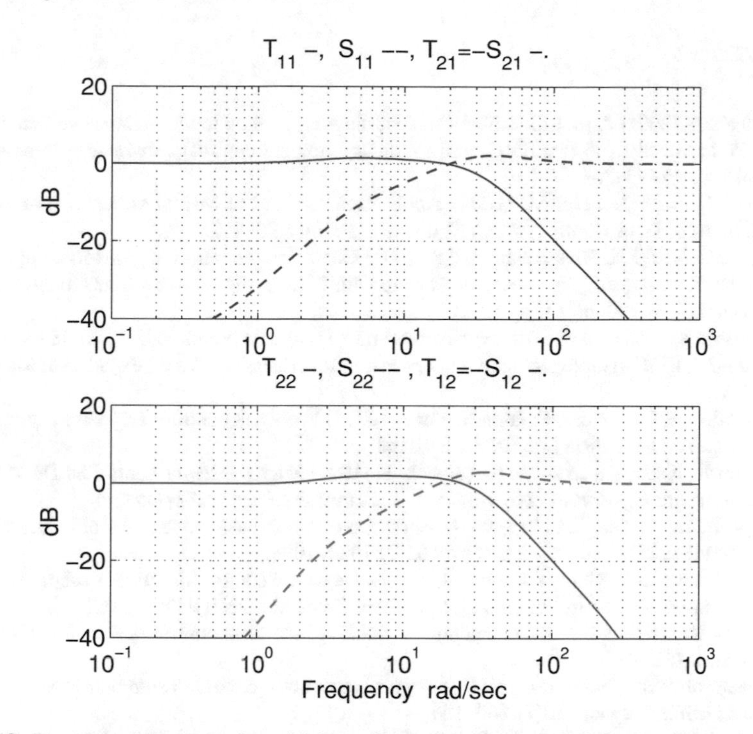

Fig. 4.19 Frequency responses of sensitivity and complementary sensitivity functions

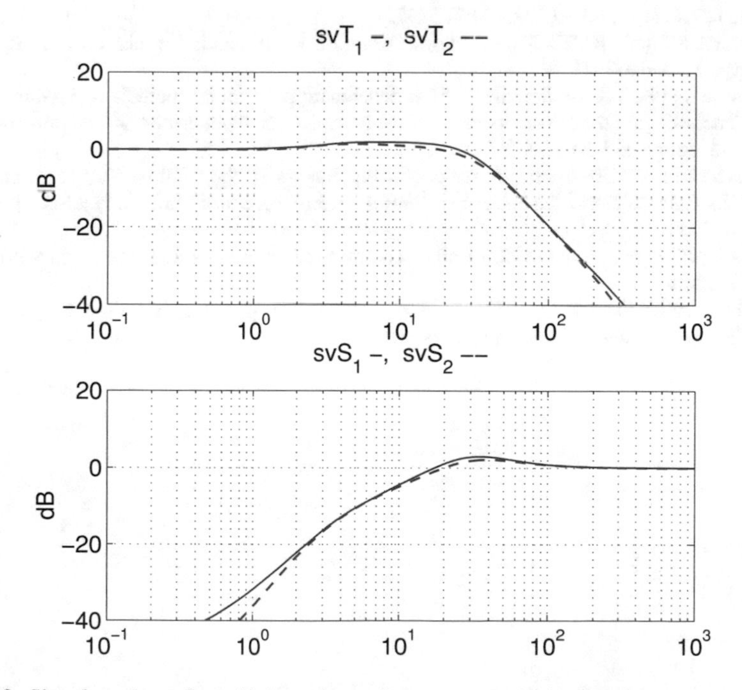

Fig. 4.20 Singular values of sensitivity and complementary sensitivity functions

References

1. Manabe S (2001) Application of coefficient diagram method to dual-control-surface missile. In: Proceedings of the 15th IFAC symposium on automatic control in aerospace, Bologna, Italy, Sept 2–7, pp 499–504
2. Kim YC, Jung DK, Hur MJ (2001) Application of CDM to MIMO systems: control of hot rolling mill. Trans Control Autom Syst Eng 3(4):2091–2096
3. Kim YC, Keel LH, Bhattacharyya SP (2003) Characteristic ratio assignment: an application to a benchmark problem. Proceedings of the 4th IFAC symposium on robust control design, Milan, Italy, vol 36, no 11, pp 55–60
4. Manabes S (1994) A low-cost inverted pendulum system for control education. In Proceedings of the 3rd IFAC symposium on advances in control education, Tokyo, Japan, August 1–2, pp 21–24
5. Franklin CF, Powell JD, Emami-Naeini A (2015) Feedback control of dynamic systems, 7th edn. Pearson Education Ltd, Essex, England
6. Ohishi K, Nakao M, Ohinishi K, Miyachi K (1993) Microprocessor-controlled DC motor for load-insensitive position servo system. IEEE Trans Ind Electron 34(1):44–49
7. Chen W, Yang J, Guo L, Li S (1993) Disturbance-observer-based control and related methods—an overview. IEEE Trans Ind Electron 63(2):1083–1096
8. Hori Y, Sawada H, Chun Y (1999) Slow resonance ratio control for vibration suppression and disturbance rejection in tortional system. IEEE Trans Ind Electron 46(1):162–168
9. Wie B, Bernstein DS (1992) Benchmark problem for robust control design. J Guid Control Dynam 15(5):1057–1059
10. Thompson PM (1995) Classical/H_2 solution for a robust control design benchmark problem. J Guid Control Dynam 18(1):160–169
11. Lipatov AV, Sokolov NI (1978) Some sufficient conditions for stability and instability of continuous linear stationary systems. Translated from Automatika i Telemekhanika 9:30–37, 1978; Auto Remote Control 39:1285–1291 (1979)
12. Alfriend KT (ed) (1992) Robust control design for a benchmark problem. J Guid Control Dynam 15(5):1057–1148
13. Manabe S (1996) A solution of the ACC benchmark problem by coefficient diagram method. In: Proceedings of the 6th workshop on astrodynamics and flight mechanics, Sagmihara, ISAS, July 15–16, pp 237–246 (1998)
14. Manabe S (1997) The application of coefficient diagram method to the ACC benchmark problem. In: Proceedings of the 2nd Asian control conference, Seoul Korea, July 22–25, pp II135–II138
15. Chiang RY, Safonov MG (1992) Robust control toolbox user's guide. The Math Works, Inc., Natick, MA
16. Safonov MG, Chiang RY (1988) CACSD using the state-space L_∞ theory—a design example. IEEE Trans Autom Control 33(5):477–479

Appendix
CDM Toolbox User's Guide for Use with Matlab

A.1 Introduction

The coefficient diagram method (CDM) is a new control design technique. In this approach, we need to calculate algebraic equations associated with Diophantine equation and draw polynomial coefficient curves on the coefficient diagram. As a computer aided design tool, the CDM toolbox (CDMTOOL) presented here enclose a set of Matlab® routines. The Matlab toolbox is in the public domain and can be downloaded from the following URL:
https://github.com/ycholkim1366/CDM-Tool
http://www.cityfujisawa.ne.jp/~manabes/CDMCAD.htm

Unzipping the downloaded file, you will have two directories, named by Work and Problem. The CDM M-files are in the "work". You may change the folder name, for example, "CDMTOOL". Hereafter, the name of this directory will be considered as "CDMTOOL". User should copy this folder to
C:\ matlab\ CDMTOOL
and then the path must be added in the Matlab environment. Otherwise, this CDM-TOOL folder may be directly copied in the Matlab toolbox directory
C:\ matlab\ toolbox
Note: It is better for you to open your working directory to create your files.

A.2 The CDM Toolbox

A number of Matlab M-files are available under the directory CDMTOOL. A summary of Matlab routines and their functions are shown below. Symbols and parameter conventions used in the CDM M-files are listed in Table A.1 and the CDM Matlab function files are listed in Tables A.2, A.3 and A.4.

S. Manabe and Y. C. Kim, *Coefficient Diagram Method for Control System Design*, Intelligent Systems, Control and Automation: Science and Engineering 99,
https://doi.org/10.1007/978-981-16-0546-8

Table A.1 CDMTOOL parameter conventions

Parameters	Description
a	Polynomial
aa	Characteristic polynomial (CP) $P(s)$
ac, bc	Denominator/numerator polynomials of the controller $G_c(s) = B_c(s)/A_c(s)$
ap, bp	Denominator/numerator polynomials of the plant $G_p(s) = B_p(s)/A_p(s)$
acp, bcp	Denominator/numerator polynomials of the open-loop function $G(s) = G_c(s)G_p(s)$, where $A_{cp}(s) = A_c(s)A_p(s)$ and $B_{cp}(s) = B_c(s)B_p(s)$
al	Ratio parameter vector in Routh Table. al$(i) := R(i)/R(i+1)$, where $R(i)$ is the element of the i-th row of the 1st column of the Routh Table
aq	Squared polynomial; $A_q(-s^2) = A(-s)A(s)$
ba	Prefilter polynomial $B_a(s)$ for command input
c	Controller
g	Stability index $\gamma = [\gamma_{n-1} \cdots \gamma_2 \gamma_1]$
gr	Desired stability index
gs	Stability limit $\gamma^* = [\gamma^*_{n-1} \cdots \gamma^*_2 \gamma^*_1]$
k	Gain
lp	Loop transfer function
mc, nc	Order of numerator/denominator polynomials of controller
mp, np	Order of numerator/denominator polynomials of plant
rr	Roots of a polynomial
t	Equivalent time constant (ETC) τ
tm	Time scale for time response
unc	Indicates whether the coefficient of the controller denominator polynomial to be normalized to 1 is the highest or lowest term
w	Diagonal entry of the weight matrix Q for LQR design

Table A.2 CDM Matlab function files: Analysis and Graphics Utility

Function	Description
`bode2`	Bode plot of two transfer functions: $G_1 = N_1/D_1$ and $G_2 = N_2/D_2$
`cdia`	Coefficient diagram of a CP and its γ, τ, γ^*
`fresp`	Frequency responses of loop transfer function $G(s)$, sensitivity function $S(s)$, and complementary sensitivity function $T(s)$ for a plant $G_p(s)$ and a controller $G_c(s)$
`rresp2/rresp3`	Step responses of two/ three transfer functions: $G_1 = N_1/D_1$, $G_2 = N_2/D_2$, and $G_3 = N_3/D_3$
`tresp`	Step responses of two transfer functions specified by a common denominator (D) and two numerators (N_1, N_2)
`tresp4`	Step responses of four transfer functions specified by a common denominator (D) and four numerators ($N_1, \ldots, N_4$)

Table A.3 CDM Matlab function files: Conversion and Computational Utility

Function	Description
`a2al`	Extract Routh Table and ratio parameters vector from the characteristic polynomial (CP)
`a2aq`	Convert a polynomial to the squared polynomial
`a2g`	Compute γ, τ, γ^*, and roots of a polynomial
`a2w`	Compute the weight matrix for LQR design from `aa` and `ap`
`al2a`	Convert the Routh ratio parameters to the corresponding polynomial
`aq2a`	Convert the squared polynomial to the original polynomial
`bpt`	Compute the break points from the CP
`c2g`	Compute the CP, γ, τ, γ^*, poles, gain/phase margins of the closed-loop system from the plant (`ap, bp`) with a controller (`ac, bc, ba`)
`c2lp`	Convert the plant and controller polynomials to the numerator/denominator polynomials of the loop transfer function
`convvm`	Convolution of a vector and a matrix
`g2a`	Compute the CP from γ and the highest/ lowest coefficients set
`g2t`	Compute possible ETC τ under the given plant model, desired stability index, and the order of controller
`gk2t`	Compute possible ETC τ under the given plant model, desired stability index, the order of controller, and a fixed steady state gain of controller numerator polynomial
`gt2a`	Compute the CP and squared polynomial from the given γ and τ

Table A.4 CDM Matlab function files: Design Utility

Function	Description
`a2c`	Find the CDM controller (`ac, bc`) from the plant (`ap, bp`) and a desired CP
`aq2c`	Find the CDM controller (`ac, bc`) from the plant (`ap, bp`) and a squared polynomial
`aqwc`	Provides the weight Q for LQR and a CDM controller from a squared polynomial and plant (`ap, bp`)
`g2c`	Find the CDM controller (`ac, bc`) from the plant (`ap, bp`), desired γ and τ, and the order of controller
`gk2c`	Provides the CDM controller obtained by adding a constraint on the steady state gain of the controller to `g2c`
`cc`	A simplified form of `c2g`
`gc`	A batch file combining `g2c` and `c2g`
`gkc`	A batch file combining `gk2c` and `c2g`

A.3 How to Use CDM Toolbox?

After the installation as illustrated in Sect. A.1, one can use the CDM toolbox function files. In this section, we demonstrate the proper syntax for entering the information to Matlab for CDM design.

a2al

Purpose:

Extract Routh Table and ratio parameters vector from the characteristic polynomial(CP).

Synopsis and Description:

```
[al, RT]=a2al(aa)
```

Input:

aa=input polynomial in descending order.

Outputs:

al= Ratio parameters generated in Routh table, defined by $\texttt{al(i)} = R(i+1)/R(i)$, where $R(i)$ is the element of the i-th row of the first column of the Routh table.

RT= Routh table

Example:

Let us consider the following polynomial:

$$P(s) = 0.25\,s^5 + s^4 + 2\,s^3 + 2\,s^2 + s + 0.2$$

which can be implemented as:

```
>>aa=[0.25 1 2 2 1 0.2];
>>[al,RT]=a2al(aa)
al =
    2.5000e-01   6.6667e-01   1.0976e+00   1.8709e+00   3.6524e+00
RT =
    2.50e-01          0     2.00e+00           0    1.00e+00          0
           0   1.00e+00            0    2.00e+00           0   2.00e-01
           0          0     1.50e+00           0    9.50e-01          0
           0          0            0  1.3667e+00           0   2.00e-01
           0          0  -2.2204e-16           0  7.3049e-01          0
           0          0            0           0           0   2.00e-01
```

a2aq

Purpose:

Convert a polynomial to the squared polynomial. For a given polynomial $P(s)$, the output is $A_q(-s^2) = P(-s)P(s)$.

Synopsis and Description:

```
aq=a2aq(aa)
```

Input:

aa=input polynomial in descending order.

Output:

aq= squared polynomial.

Example:

Let us consider the following polynomial:

$$P(s) = s^4 + 2s^3 + 2s^2 + s + 0.2$$

which can be implemented as

```
>> aa=[1 2 2 1 0.2];
>> aq=a2aq(aa)
aq =
    1.0000e+00  0  4.0000e-01  2.0000e-01  4.0000e-02
```

The above vector aq can be represented in the form, with $x = -s^2$

$$A_q(x) = x^4 + 0.4x^2 + 0.2x + 0.04 \quad \Rightarrow \quad A_q(-s^2) = s^8 + 0.4s^4 - 0.2s^2 + 0.04.$$

a2c

Purpose:

Find the CDM controller (ac, bc) so that a desired CP is assigned for the given plant (ap, bp).

Synopsis and Description:

```
[bc,ac,aa,g,tau,gs,rr]=a2c(ap,bp,ar,unc)
```

Inputs:

`bp,ap`= numerator and denominator polynomials of the plant.
`ar`= a reference polynomial for the desired closed-loop system.
`unc`=indicates whether the coefficient of the controller denominator polynomial to be normalized to 1 is the highest or lowest term, which is normally 0 or `nc`.

Outputs:

`bc,ac`= numerator and denominator polynomials of the controller.
`aa`= characteristic polynomial of the closed loop system with a designed controller.
`g`= stability index of `aa`.
`tau`= equivalent time constant of `aa`.
`gs`= stability limit of `aa`.
`rr`= roots of characteristic polynomial `aa`.

Example:

Consider the following plant and the reference characteristic polynomial:

$$G_p(s) = \frac{1}{s^4+s^2}$$

$$P_r(s) = 2^{-9}s^7 + 2^{-5}s^6 + 2^{-2}s^5 + s^4 + 2s^3 + 2s^2 + s + 0.2$$

that will be formed as,

```
>> ap=[1 0 1 0 0]; bp=[1]; ar=[2^-9 2^-5 2^-2 1 2 2 1 0.2];
>> [bc,ac,aa,g,tau,gs,rr]=a2c(ap,bp,ar,0)

bc =
    1.8085e+00   1.0645e+00   1.0323e+00   2.0645e-01
ac =
    2.0161e-03   3.2258e-02   2.5605e-01   1.0000e+00
aa =
  2.016e-3  3.226e-2  2.581e-1  1.032  2.065  2.065  1.032  2.065e-1
g =
    2.00e+00  2.00e+00  2.00e+00  2.00e+00  2.00e+00  2.50e+00
tau = 5
gs =
    5.00e-01  1.00e+00  1.00e+00  1.00e+00  9.00e-01  5.00e-01
rr =
    -4.4502e+00 ± 5.2133e+00i
     -2.9270e+00 + 0.0000e+00i
     -2.2981e+00 + 0.0000e+00i
      -6.4216e-01 ± 3.6956e-01i
      -5.9026e-01 + 0.0000e+00i
```

a2g

Purpose:

Compute γ, τ, γ^*, and roots of the characteristic polynomial $P(s)$.

Synopsis and Description:

```
[g,tau,gs,rr]=a2g(aa)
```

Input:

aa= coefficients of the characteristic polynomial $P(s)$.

Outputs:

g= stability index of aa.
tau= equivalent time constant of aa.
gs= stability limit of aa.
rr= roots of characteristic polynomial aa.

Example:

Consider the following characteristic polynomial:

$$P(s) = s^4 + 2s^3 + 2s^2 + s + 0.2$$

will be formed as,

```
>> aa=[1 2 2 1 0.2];
>> [g,tau,gs,rr]=a2g(aa)

g =
    2.0000  2.0000  2.5000
tau = 5
gs =
    0.5000  0.9000  0.5000
rr =
    -0.5000 ±0.6882i
    -0.5000 ±0.1625i
```

a2w

Purpose:

Compute the weight matrix for LQR design from the characteristic polynomial and plant denominator polynomial.

Synopsis and Description:

```
[aq,apq,qq]=a2w(aa,ap)
```

Inputs:

ap= denominator polynomial $A_p(s)$ of the plant .
aa= characteristic polynomial $P(s)$.

Outputs:

aq= squared polynomial of aa, defined as $A_q(-s^2) = P(-s)P(s)$.
apq= squared polynomial of ap, defined as $A_{pq}(-s^2) = A_p(-s)A_p(s)$.
qq= [q11 q22 ⋯ qnn], where q_{ii} is the diagonal element of the weight matrix of Q.

Example:

Consider the following plant denominator and characteristic polynomials:

$$A_p(s) = s^4 + s^2$$
$$P(s) = s^4 + 2s^3 + 2s^2 + s + 0.2$$

They will be formed as

```
>> ap=[1 0 1 0 0]; aa=[1 2 2 1 0.2];
>> [aq,apq,qq]=a2w(aa,ap)
aq =
    1.0000   0   0.4000   0.2000   0.0400
apq =
    1  -2  1  0  0
qq =
    2.0000  -0.6000  0.2000  0.0400
```

The above vector aq can be represented in the form, with $x = -s^2$

$$A_q(x) = x^4 + 0.4x^2 + 0.2x + 0.04 \quad \Rightarrow \quad A_q(-s^2) = s^8 + 0.4s^4 - 0.2\,s^2 + 0.04.$$

al2a

Purpose:

Convert the Routh ratio parameters to the corresponding characteristic polynomial.

Synopsis and Description:

```
[aa,RT]=al2a(al)
```

Input:

`al=` ratio parameter vector in Routh Table. `al`$(i) = R(i)/R(i+1)$, where $R(i)$ is the i-th element of the first column of the Routh Table.

Outputs:

`aa=` characteristic polynomial corresponding to the given `al`.
`RT=` Routh table of `aa`.

Example:

Consider the following characteristic polynomial:

$$P(s) = 0.25s^5 + s^4 + 2s^3 + 2s^2 + s + 0.2$$

It is easy to know by using M-function `a2al` that Routh ratio parameter vector of $P(s)$ above is given as `al`=[0.25 0.66667 1.0976 1.8709 3.6524].
Then

```
>>[aa,RT]=al2a(al)
aa =
    1   4   8   8   4   0.8
RT =
    1.0000        0   8.0000        0   4.0000        0
         0   4.0000        0   8.0000        0   0.8000
         0        0   6.0000        0   3.8000        0
         0        0        0   5.4667        0   0.8000
         0        0        0        0   2.9220        0
         0        0        0        0        0   0.8000
```

aq2a

Purpose:

Convert the squared polynomial $A_q(-s^2)$ to the original polynomial $P(s)$ and its γ, τ, γ^*.

Synopsis and Description:

```
[aa,g,tau,gs,rr]=aq2a(aq)
```

Input:

`aq=` squared polynomial $A_q(-s^2)$.

Outputs:

`aa=` characteristic polynomial $P(s)$.
`g=` stability index of `aa`.
`tau=` equivalent time constant of `aa`.

gs= stability limit of aa.
rr= roots of characteristic polynomial $P(s)$.

Example:

Consider the following squares polynomial:

$$A_q(-s^2) = s^8 + 0.4s^4 - 0.2s^2 + 0.04, \quad \text{or } A_q(x) = x^4 + 0.4x^2 + 0.2x + 0.04, \ (x = -s^2)$$

that will be formed as,

```
>> aq=[1 0 0.4 0.2 0.04];
>> [aa,g,tau,gs,rr]=aq2a(aq)
aa =
    1.0   2.0   2.0   1.0   0.2
g =
    2.00   2.00   2.5
tau = 5.0
gs =
    0.50   0.90   0.50
rr =
    -0.50 ± 0.6882i
     -0.50 ± 0.1625
```

aq2c

Purpose:

Find the CDM controller (ac,bc) from the given plant (ap,bp) and a squared polynomial (aq).

Synopsis and Description:

```
[bc,ac,aa,g,tau,gs,rr]=aq2c(ap,bp,aq,unc)
```

Inputs:

bp,ap= numerator and denominator polynomials of the plant.
aq= a squared polynomial of the characteristic polynomial.
unc=indicates whether the coefficient of the controller denominator polynomial to be normalized to 1 is the highest or lowest term, which is normally 0 or nc.

Outputs:

bc,ac= numerator and denominator polynomials of the controller.
aa= characteristic polynomial of the closed loop system with a designed controller.
g= stability index of aa.

`tau`= equivalent time constant of `aa`.
`gs`= stability limit of `aa`.
`rr`= roots of characteristic polynomial `aa`.

Example:

Consider the following plant and a squared polynomial of the characteristic polynomial:

$$G_p(s) = \frac{1}{s^4+s^2}$$

$$A_q(-s^2) = -s^{14} + 1, \quad \text{or} \quad A_q(x) = x^7 + 1, \text{(where } x = -s^2)$$

that will be formed as

```
>> ap=[1 0 1 0 0]; bp=[1]; aq=[1 0 0 0 0 0 0 1];
>> [bc,ac,aa,g,tau,gs,rr]=aq2c(ap,bp,aq,0)
bc =
    5.4407e-01   -8.7957e-15   4.4504e-01   9.9031e-02
ac =
    9.9031e-02   4.4504e-01   9.0097e-01   1.0000e+00
aa =
    9.903e-02  4.450e-01  1.000  1.445  1.445  1.000  4.450e-10  9.903e-02
g =
    2.000   1.555   1.445   1.445   1.555   2.000
tau = 4.494
gs =
    0.6431   1.1920   1.3351   1.3351   1.1920   0.6431
rr =
    -2.2252e-01 ± 9.7493e-01i
    -6.2349e-01 ± 7.8183e-01i
   -1.0000e+00 + 0.0000e+00i
    -9.0097e-01 ± 4.3388e-01i
```

aqwc

Purpose:

From the given plant (`ap,bp`) and a squared polynomial (`aq`), this M-file provides the weights for LQR and a CDM controller (`ac,bc,ba`) by using `a2c` and `aq2a`. Also, it shows the frequency responses as well as the step responses of the overall system with the resulting controller by using `c2g`.

Synopsis and Description:

```
aqwc
```

Initial information:

bp, ap= numerator and denominator polynomials of the plant.
aq= a squared polynomial of the characteristic polynomial.

Outputs:

apq= squared polynomial of (ap), $A_{pq}(-s^2) = A_p(-s)A_p(s)$.
bpq= squared polynomial of (bp), $B_{pq}(-s^2) = B_p(-s)B_p(s)$.
qu= weight polynomial for the control input u in LQR design.
qy= weight polynomial for the system output y in LQR design.
ba= feedforward polynomial of the CDM controller.
bc, ac= numerator and denominator polynomials of the CDM controller.
aa= characteristic polynomial of the closed loop system
g= stability index of aa.
tau= equivalent time constant of aa.
gs= stability limit of aa.
rr= roots of characteristic polynomial aa.
pmgm= phase and gain margins of the loop transfer function
wpmgm= phase and gain crossover frequencies of the loop transfer function

In LQR design, we may use $R = \text{qu}(1)$ and $Q = diag([\text{qu}(2) \ \cdots \ \text{qu}(nc+1) \ \text{qy}(1) \ \cdots \ \text{qy}(np)])$.

Example:

Consider the following plant and a squared polynomial of the characteristic polynomial in Matlab scripts.

```
>> ap=[0.25  1.25  1  0]; bp=[0.1  1];
>> aq=[0.13598  13.772  35.757  221.25  470.38  400];
>> aqwc
apq =
    6.2500e-02   1.0625e+00   1.0000e+00   0
bpq =
    1.0000e-02   1.0000e+00
qu =
    2.1757e+00   1.8337e+02   -3.1087e+03
qy =
    3.3052e+03   3.5751e+03   4.0000e+02
ba =2.0000e+01

bc =
    2.6491e+01   4.5499e+01   2.0000e+01
ac =
    1.4750e+00   1.4750e+01   9.9722e-01
aa =
    0.3688   5.5313   22.811   47.037   48.496   20.00
g =
```

```
    3.6372    2.00    2.00    2.50
tau = 2.4248
pmgm =
    45.761    1.0179e+04
wpmgm =
    1.7715e+00    2.7051e+02
```

The rest of the outputs is Bode plot of $G(s)$, $T(s)$, $S(s)$, step responses, and pole map of the closed-loop system, as shown in Fig. A.1.

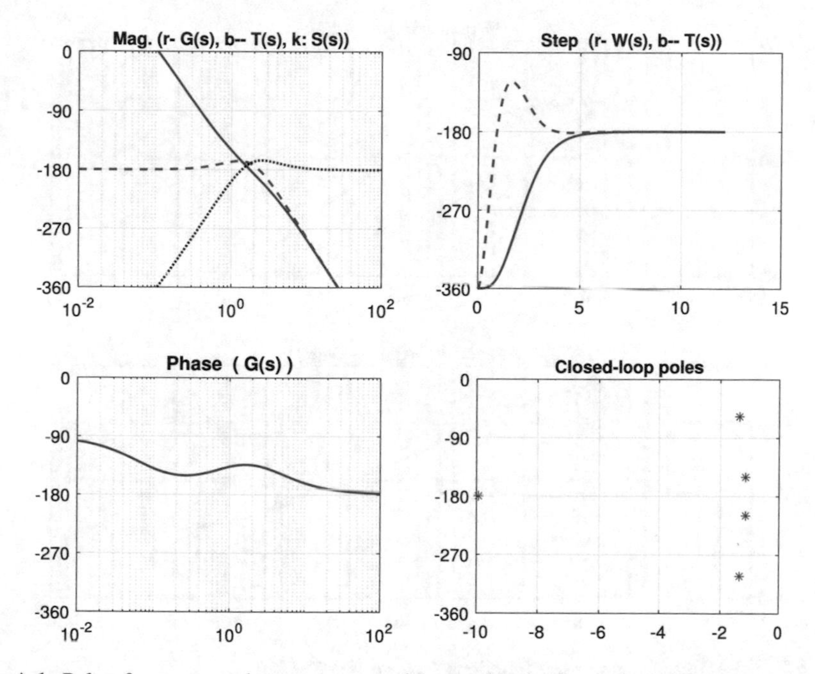

Fig. A.1 Poles, frequency and step responses of the resulting systems obtained by `aqwc`

bode2

Purpose:

This M-file provides a Bode plot for two transfer functions specified by $G_1(s) = N_1(s)/D_1(s)$ and $G_2(s) = N_2(s)/D_2(s)$.

Synopsis and Description:

```
bode2(num1,den1,num2,den2)
```

Inputs:

`num1, den1`= numerator and denominator polynomials of $G_1(s)$.

`num2, den2`= numerator and denominator polynomials of $G_2(s)$.

Example:

Consider the following two transfer functions in Matlab scripts.

```
>> num1=[1.5  1];  den1=[2.5  14.125  18.375  7.75  1];
>> num2=[1];  den2=[1.6667  8.6528  6.25  1];
>> bode2(num1,den1,num2,den2)
```

The output is the plot in Fig. A.2.

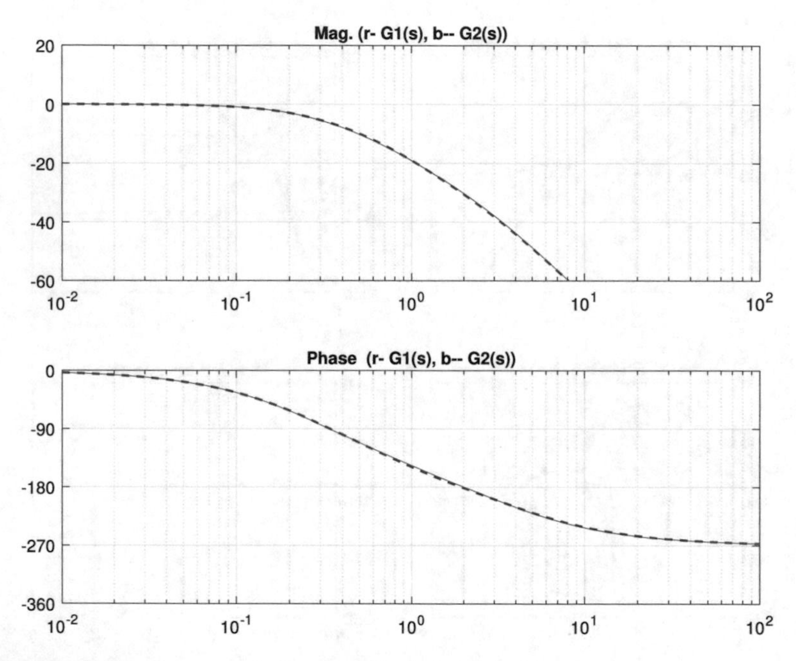

Fig. A.2 Bode plot of two transfer functions $G_1(j\omega)$ and $G_2(j\omega)$

bpt

Purpose:

Determine the pseudo-break points of a real polynomial.

Synopsis and Description:

```
[bpp]=bpt(aa)
```

Input:

`aa`= a real polynomial, $P(s) = a_n s^n + a_{n-1} s^{n-1} + \cdots + a_0$.

Output:

bpp= a vector of pseudo-break points of $P(s)$, defined as bpp$(i) = a_{n-i}/a_{n+1-i}$.

Example:

Consider the following polynomial:

$$P(s) = s^4 + 2s^3 + 2s^2 + s + 0.2$$

that will be formed as

```
>> aa=[1  2  2  1  0.2];
>> [bpp]=bpt(aa)

bpp =
    2.0   1.0   0.5   0.2
```

c2g

Purpose:

Compute the CP, γ, τ, γ^*, poles, gain/ phase margins of the closed-loop system from the plant (ap, bp) with a designed controller (ac, bc, ba).

Synopsis and Description:

```
[aa,g,tau,gs,rr,pmgm,wpmgm]=c2g(ap,bp,ac,bc,ba,tm)
```

Inputs:

bp,ap= numerator and denominator polynomials of the plant.
bc,ac= numerator and denominator polynomials of the controller.
ba= numerator polynomial for the command input.
tm= the time scale of the time response figure, which is normally set to 1.0 or 0.5.

Outputs:

aa= characteristic polynomial of the closed loop system with a controller.
g= stability index of aa.
tau= equivalent time constant of aa.
gs= stability limit of aa.
rr= roots of characteristic polynomial aa.
pmgm= phase and gain margins of the loop transfer function
wpmgm= phase and gain crossover frequencies of the loop transfer function

Example:

Consider the following plant and the controller in two-parameter configuration:

$$G_p(s) = \frac{B_p(s)}{A_p(s)} = \frac{1}{s^3+s}$$

$$A_c(s) = 0.071794s^2 + 0.3923s + 1, \quad B_c(s) = 1.0718s^2 + 0.27322,$$
$$B_a(s) = 0.27322,$$

that will be formed as

```
>> ap=[1 0 1 0];  bp=[1];
>> ac=[0.071794 0.3923 1];  bc=[1.0718 0 0.27322];
>> ba=[0.27322];  tm=1.0
>> [aa,g,tau,gs,rr,pmgm,wpmgm]=c2g(ap,bp,ac,bc,ba,tm)
aa =
    7.1794e-02  0.3923  1.0718  1.4641  1.0000  2.7322e-01
g =
    2.000   2.000   2.000   2.4999
tau = 3.6601
gs =
    0.500   0.9999   9.0002e-01   0.500
rr =
    -1.5182e+00 ± 1.7480e+00i
     -8.2521e-01 ± 4.8197e-01i
     -7.7733e-01 + 0.0000e+00i
pmgm =
    5.3156e+01   4.8205e+00
wpmgm =
    1.5714e+00   3.7321e+00
```

The rest of the outputs is frequency response of $G(s)$, $T(s)$, $S(s)$, step responses, and pole map of the closed-loop system, as shown in Fig. A.3.

c2lp

Purpose:

Convert the plant $G_p(s) = B_p(s)/A_p(s)$ and controller $G_c(s) = B_c(s)/A_c(s)$ to the numerator and denominator polynomials of the loop transfer function $G(s) = G_c(s)G_p(S)$.

Synopsis and Description:

```
[bcp,acp,aa]=c2lp(ap,bp,ac,bc)
```

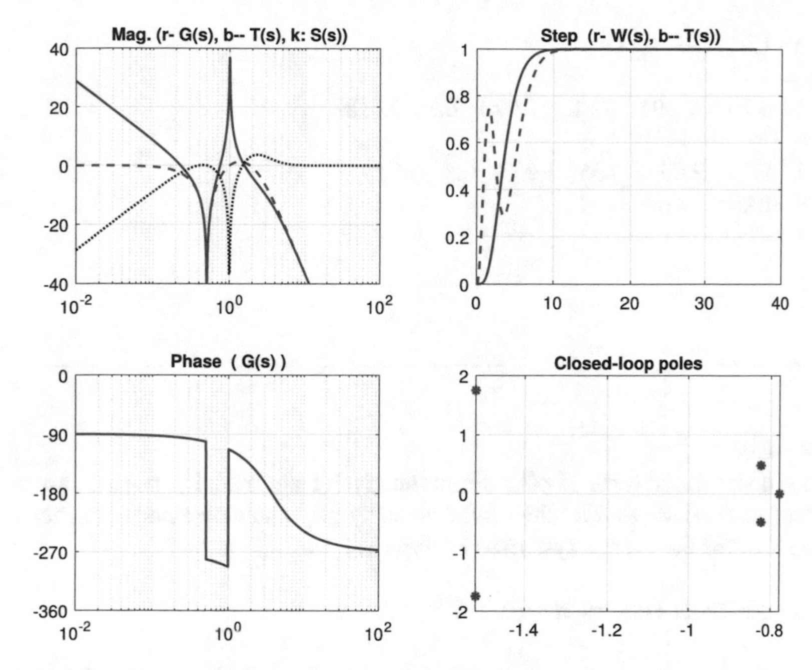

Fig. A.3 Poles, frequency and step responses of the resulting systems provided by `c2g`

Inputs:

`bp,ap`= numerator and denominator polynomials of the plant.
`bc,ac`= numerator and denominator polynomials of the controller.

Outputs:

`bcp,acp`=numerator and denominator polynomials of the loop transfer function, where $B_{cp}(s) = B_p(s)B_c(s)$ and $A_{cp}(s) = A_p(s)A_c(s)$.
`aa`= characteristic polynomial of the closed loop system.

Example:

Consider the following single input-two output plant and its controller:

$$A_p(s) = s^4 - 3s^2 + 2, \quad B_p(s) = \begin{bmatrix} 2s^3 - 2s \\ s^3 - 2s \end{bmatrix}$$
$$A_c(s) = s^2 + 0.15s + 0.009, \quad B_c(s) = \begin{bmatrix} 8.5s^2 + 12s \\ -12s^2 - 12s \end{bmatrix},$$

that will be formed as

```
>> ap=[1 0 -3 0 2];  bp=[2 0 -2 0; 1 0 -2 0];
>> ac=[1  0.15  0.009 ];  bc=[8.5  12  0; -12  -12  0];
>> [bcp,acp,aa]=c2lp(ap,bp,ac,bc)
bcp =
```

```
    5  12  7  0  0  0
acp =
    1  0.15  -2.991  -0.45  1.973  0.3  0.018
aa =
    1  5.15  9.009  6.55  1.973  0.3  0.018
```

See also: `lp`

cc

Purpose:

This is a simplified form of `c2g`. From the given plant model (`ap`, `bp`) and a controller (`ac`, `bc`), it provides the results given by `c2g` and a command prefilter polynomial `ba` for the overall system to be Type 1.

Synopsis and Description:

```
cc
```

Initial information:

`bp, ap`= numerator and denominator polynomials of the plant.
`bc, ac`= numerator and denominator polynomials of the controller.

Outputs:

`ba`= feedforward polynomial of the CDM controller.
`aa`= characteristic polynomial of the closed loop system
`g`= stability index of `aa`.
`tau`= equivalent time constant of `aa`.
`gs`= stability limit of `aa`.
`rr`= roots of characteristic polynomial `aa`.
`pmgm`= phase and gain margins of the loop transfer function
`wpmgm`= phase and gain crossover frequencies of the loop transfer function

Example:

Consider a single input double output plant and a controller in Matlab scripts as follows:

```
>> ap=[1  0  2  0];  bp=[0  0  1;  1  0  1];
>> ac=[0.1  1];  bc=[10.5  7.7;  0.0  4.8];
>> cc
ba = 12.5
aa =
```

```
    0.1  1  5  12.5  12.5
g =
    2.0  2.0  2.5
tau = 1.0
gs =
    0.5  0.9  0.5
rr =
    -2.5 ± 3.441i
    -2.5± 0.8123i
pmgm =
    37.083  0.12696
wpmgm =
    4.746  1.8258
```

The rest of the outputs is frequency response of $G(s)$, $T(s)$, $S(s)$, step responses, and pole map of the closed-loop system, as shown in Fig. A.4.

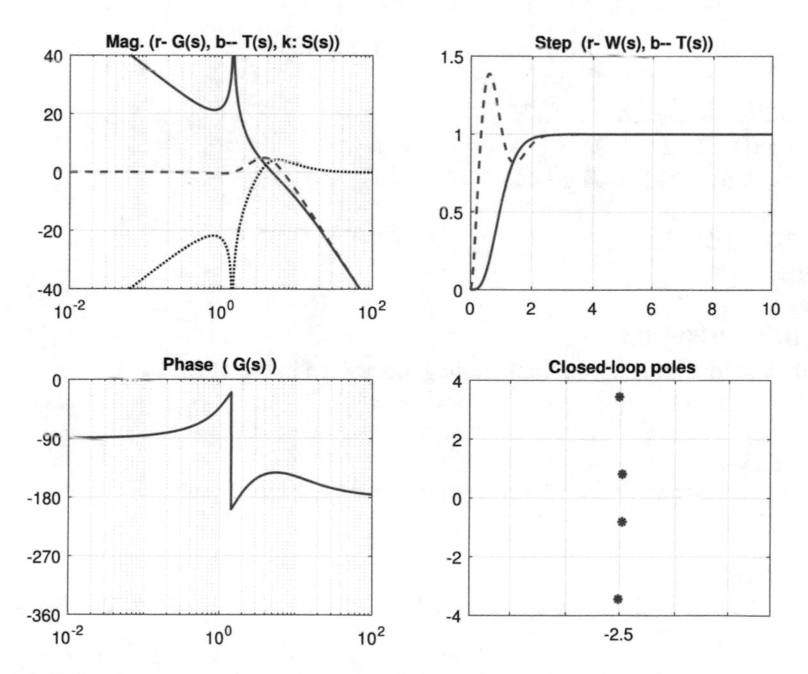

Fig. A.4 Poles, frequency and step responses of the resulting systems provided by `cc`

cdia

Purpose:

Draw the coefficient diagram for a given CP, $P(s)$ and provides its γ, τ, and γ^*.

Synopsis and Description:

```
[g,tau,gs]=cdia(aa)
```

Input:

aa= characteristic polynomial $P(s)$.

Outputs:

g= stability index of aa.
tau= equivalent time constant of aa.
gs= stability limit of aa.

Example:

Consider the following characteristic polynomial:

$$P(s) = s^4 + 2s^3 + 2s^2 + s + 0.2,$$

that will be formed as

```
>> aa=[1  2  2  1  0.2];
>> [g,tau,gs]=cdia(aa)
g =
   2.0   2.0   2.5
tau =    5
gs =
   0.50   0.90   0.50
```

The other output is the coefficient diagram shown in Fig. A.5.

convvm

Purpose:

Obtain the convolution of a row vector $V \in \mathcal{R}^{n_v}$ and a matrix $M \in \mathcal{R}^{m_m \mathrm{x} n_m}$, which results in $V_M \in \mathcal{R}^{m_m \mathrm{x}(\mathrm{n_v}+\mathrm{n_m}-1)}$.

Synopsis and Description:

```
[vm]=convvm(vv,mm)
```

Inputs:

vv= a row vector representing a polynomial $V(s)$.
mm= a matrix representing a polynomial matrix $M(s) = [M_{11}(s)\ M_{21}(s)\ \cdots\ M_{m1}(s)]^T$.

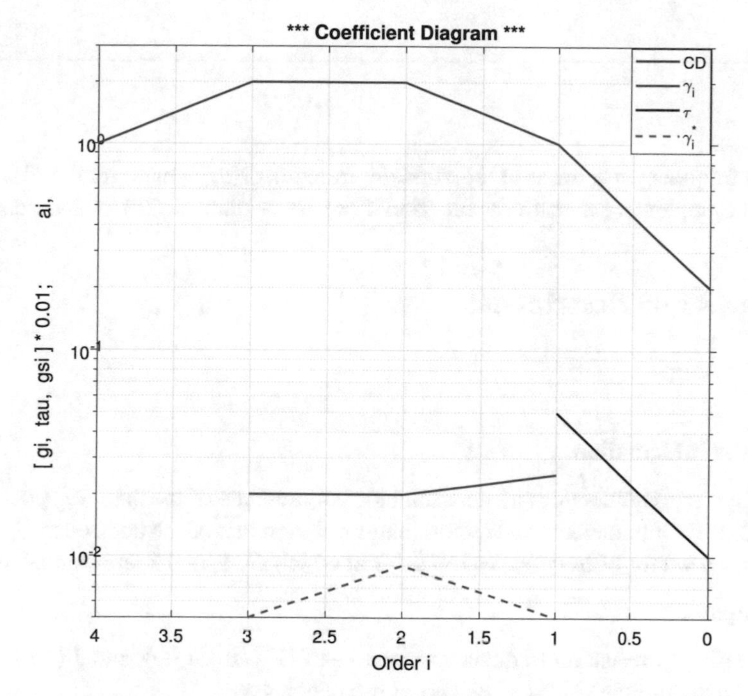

Fig. A.5 Coefficient diagram of the characteristic polynomial in the Example

Output:

vm= a matrix representing a polynomial matrix $V_M(s) = [V(s) * M_{11}(s)\ V(s) * M_{21}(s)\ \cdots\ V(s) * M_{m1}(s)]^T$.

Example:

Consider the following two polynomials:

$$V(s) = s + 1, \qquad M(s) = \begin{bmatrix} 3s^2 + 2s + 1 \\ s^2 + s \end{bmatrix}.$$

Let us find the convolution of $V(s)$ and $M(s)$.

```
>> vv=[1  1]; mm=[3  2  1;  1  1  0];
>> [vm]=convvm(vv,mm)
vm =
    3  5  3  1
    1  2  1  0
```

The vector vm represents

$$V_M(s) = V(s) * M(s) = \begin{bmatrix} 3s^3 + 5s^2 + 3s + 1 \\ s^3 + 2s^2 + s \end{bmatrix}.$$

fresp

Purpose:

Obtain frequency responses of loop transfer function $G(s)$, sensitivity function $S(s)$, and complementary sensitivity function $T(s)$ for a plant $G_p(s)$ and a controller $G_c(s)$.

Synopsis and Description:

```
fresp
```

Initial information:

`bp, ap`= numerator and denominator polynomials of the plant $G_p(s)$.
`bc, ac`= numerator and denominator polynomials of the controller $G_c(s)$.
w= a vector of frequencies in rad/s at which G, S, and T are evaluated.

Outputs:

a plot representing frequency responses of $G(j\omega)$, $S(j\omega)$, and $T(j\omega)$.
`G`= magnitude values of $G(j\omega)$ at frequencies w.
`Gphase`= phase values of $S(j\omega)$ at frequencies w.
`T`= magnitude values of $T(j\omega)$ at frequencies w.
`S`= magnitude values of $S(j\omega)$ at frequencies w.

Example:

Consider a single input double output plant and a 2x1 controller in Matlab scripts as follows:

```
>> ap=[1 0 2 0]; bp=[0 0 1; 1 0 1];
>> ac=[0.1 1]; bc=[10.5 7.7; 0.0 4.8];
>> w= [0.5 1 2];
>> fresp
w =
    0.5  1  2
G =
    1.4222e+01   1.2956e+01   5.4037e+0
Gphase =
    -4.2794e+02   -4.0196e+02   -1.7361e+02
T =
    9.7232e-01   9.4460e-01   1.2250e+00
S =
    6.8366e-02   7.2908e-02   2.2669e-01
```

The rest of the outputs is frequency response of $G(s)$, $T(s)$, and $S(s)$ shown in Fig. A.6.

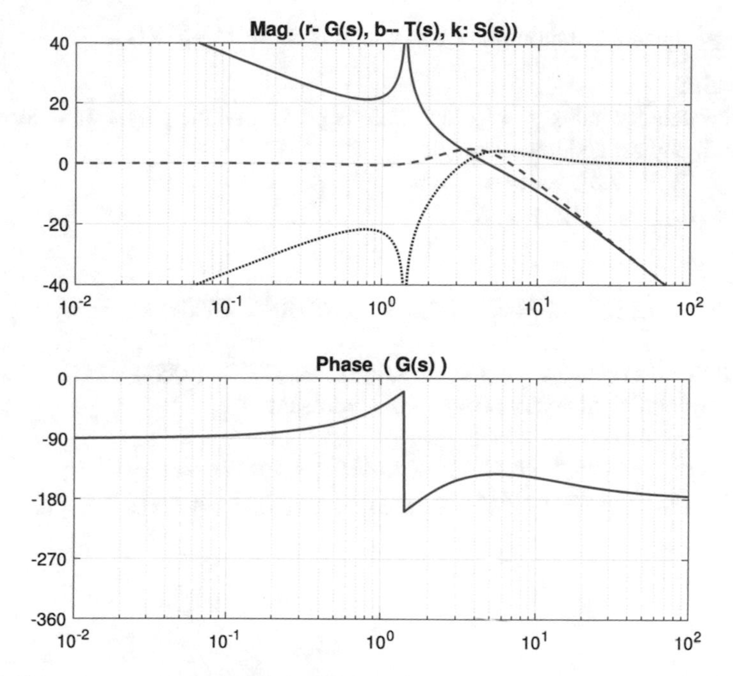

Fig. A.6 Frequency responses of $G(j\omega)$, $S(j\omega)$, and $T(j\omega)$ using `fresp`

g2a

Purpose:

Obtain the polynomial and the squared polynomial corresponding to the specified stability index, in which the coefficients of the highest and lowest orders are set to the given values.

Synopsis and Description:

```
[aa,aq]=g2a(g,an,a0)
```

Inputs:

`g`= stability index
`an,a0`= coefficients of the highest/ lowest terms of polynomial.

Outputs:

`aa`= polynomial $P(s)$ corresponding to the specified (`g, an, a0`).

aq= squared polynomial of aa, which is $A_q = P(-s)P(s)$.

Example:

When the stability index $\gamma = [2\ \ 2\ \ 2\ \ 2.5]$ is given, find the polynomial normalized at the both highest and lowest orders.

```
>> g=[2  2  2  2.5];  an=1;  a0=1;
>> [aa,aq]=g2a(g,an,a0)

aa =
    1.0000    4.1826    8.7469    9.1461    4.7818    1.0000
aq =
    1.0000    1.0658e-14    9.5635    8.3651    4.5731    1.0000
```

The above coefficient vectors aa and aq represent

$$P(s) = s^5 + 4.1826s^4 + 8.7469s^3 + 9.1461s^2 + 4.7818s + 1,$$
$$A_q(x) = x^5 + 1.0658\text{x}10^{-14}x^4 + 9.5635x^3 + 8.3651x^2 + 4.5731x + 1, \quad (x = -s^2).$$

g2c

Purpose:

Find a CDM controller (ac, bc) for a given plant (ap, bp) under the design parameters: desired stability index γ, equivalent time constant τ, and the order of controller.

Synopsis and Description:

```
[bc,ac,aa,g,tau,gs,rr]=g2c(ap,bp,nc,mc,gr,t,unc)
```

Inputs:

bp, ap= numerator and denominator polynomials of the plant.
mc, nc= numerator and denominator orders of the controller.
gr= a vector of the desired stability index.
t= equivalent time constant τ. If t is a vector, the left end value specifies a higher-order equivalent time constant.
unc=indicates whether the coefficient of the controller denominator polynomial to be normalized to 1 is the highest or lowest term, which is normally 0 or nc.

Outputs:

bc, ac= numerator and denominator polynomials of the CDM controller.
aa= characteristic polynomial of the closed loop system

`g`= stability index of the resulting `aa`.
`tau`= equivalent time constant of `aa`.
`gs`= stability limit of `aa`.
`rr`= roots of characteristic polynomial `aa`.

Example:

Consider the following third order plant:

$$G_p(s) = \frac{s + 0.1}{s^3}.$$

Suppose that we want to find a first-order controller $G_c(s) = (k_1(s) + k_0)/(l_1 s + 1)$ with design parameters $\gamma = [2\ \ 2\ \ 2.5]$ and $\tau = 2.0$.

```
>> ap=[1 0 0 0]; bp=[1 0.1];
>> gr=[2 2 2.5]; nc=1; mc=1; unc=0;
>> [bc,ac,aa,g,tau,gs,rr]=g2c(ap,bp,nc,mc,gr,t,unc)

bc =
    1.0000e+00  4.0000e-01
ac =
    5.0000e-01  1.0000e+00
aa =
    5.0000e-01  1.0000e+00  1.0000e+00  5.0000e-01  4.0000e-02
g =
    2.00  2.00  6.25
tau =
    2.0000e+00  1.2500e+01
gs =
    5.0000e-01  6.6000e-01  5.0000e-01
rr =
    -5.0000e-01 ± 8.1382e-01i
    -9.0288e-01 + 0.0000e+00i
    -9.7122e-02 + 0.0000e+00i
```

g2t

Purpose:

Find possible equivalent time constants τ for a given plant (`ap,bp`) under the conditions of desired stability index γ and the order of controller.

Synopsis and Description:

```
tau=g2t(ap,bp,nc,mc,gr)
```

Inputs:

`bp,ap`= numerator and denominator polynomials of the plant.
`mc,nc`= numerator and denominator orders of the controller.
`gr`= a vector of the desired stability index.

Outputs:

`tau`= possible equivalent time constants.

Example:

Consider the following third order plant.

$$G_p(s) = \frac{1}{0.25s^3 + 1.25s^2 + s}.$$

Let us find possible equivalent time constants under the condition of the desired stability index of $\gamma = [4\ 2\ 2.5]$ when a first-order controller $G_c(s) = (k_1 s + k_0)/(l_1 s + 1)$ is considered.

```
>> ap=[0.25  1.25  1  0];   bp=[1];
>> gr=[4  2  2.5];   nc=1;   mc=1;
>> tau=g2t(ap,bp,nc,mc,gr)

tau =
            0
            0
   3.3333e+00
   1.4286e+00
```

gc

Purpose:

A batch file combining `g2c` and `c2g`. It provides various design results including CDM controller and their time/frequency responses.

Synopsis and Description:

```
gc
```

Initial information:

`bp,ap`= numerator and denominator polynomials of the plant $G_p(s)$.
`mc,nc`= numerator and denominator orders of the controller $G_c(s)$.
`gr`= a reference stability index.
`t`= a desired equivalent time constant

tm= time scale for time response, which is normally set to 1 or 0.5.

Outputs:

a plot representing time and frequency responses of $G(s)$, $T(s)$, and $S(s)$.
ba, bc, ac= polynomials of the designed CDM controller.
aa= characteristic polynomial of the closed loop system
g= stability index of aa.
tau= equivalent time constant of aa.
gs= stability limit of aa.
rr= roots of characteristic polynomial aa.
pmgm= phase and gain margins of the loop transfer function
wpmgm= phase and gain crossover frequencies of the loop transfer function

Example:

Consider a single input double output plant and a 2x1 controller in Matlab scripts as follows:

```
>> ap=[1 0 2 0]; bp=[0 0 1; 1 0 1];
>> gr=[2 2 2.5]; nc=1 mc=[1; 0]; t= 1; tm= 1
>> gc

ba =1.2500e+01

bc =
    1.0500e+01   7.70
             0   4.80
ac =
    1.0000e-01   1.0000
aa =
    1.0000e-01   1.00   5.00   1.2500e+01   1.2500e+01
g =
    2.00   2.00   2.50
tau =1


gs =
    0.5   0.9   0.5
rr =
    -2.5000e+00 ± 3.4410e+00i
    -2.5000e+00 ± 8.1230e-01i
pmgm =
    3.7083e+01   1.2696e-01
wpmgm =
    4.7460e+00   1.8258e+00
```

The rest of the outputs is frequency response of $G(s)$, $T(s)$, $S(s)$, step responses, and pole map of the closed-loop system, as shown in Fig. A.7.

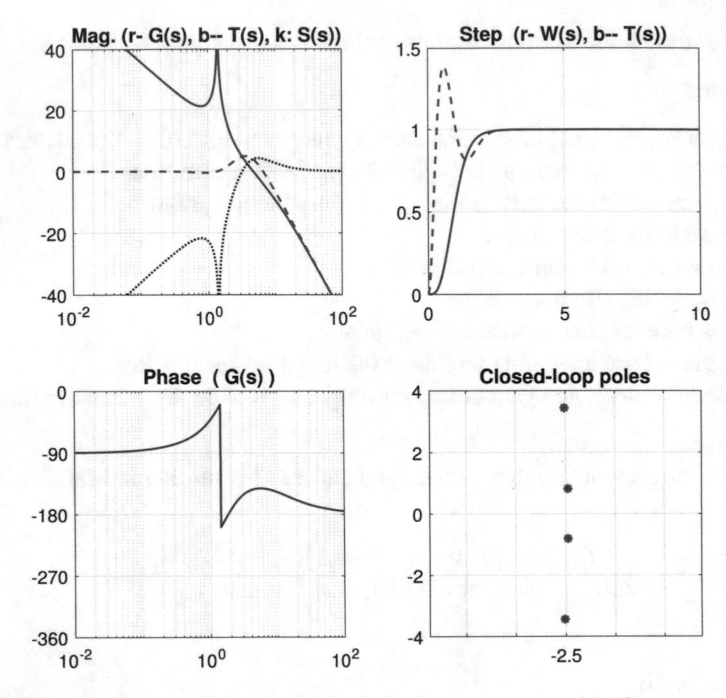

Fig. A.7 Poles, frequency and step responses of resulting systems provided by `gc`

gk2c

Purpose:

Find a CDM controller (`ac,bc`) for a given plant (`ap,bp`) by adding a constraint on the steady state gain of controller to `g2c`.

Synopsis and Description:

```
[bc,ac,aa,g,tau,gs,rr]=gk2c(ap,bp,nc,mc,gr,t,k0)
```

Inputs:

`bp,ap`= numerator and denominator polynomials of the plant.
`mc,nc`= numerator and denominator orders of the controller.
`gr`= a vector of the desired stability index.
`t`= equivalent time constant τ.
`k0`= a steady state gain of numerator polynomial of the controller.

Outputs:

`bc,ac`= numerator and denominator polynomials of the CDM controller.
`aa`= characteristic polynomial of the closed loop system
`g`= stability index of the resulting `aa`.
`tau`= equivalent time constant of `aa`.
`gs`= stability limit of `aa`.
`rr`= roots of characteristic polynomial `aa`.

Example:

Consider the following third order plant:

$$G_p(s) = \frac{1}{0.25s^3 + 1.25s^2 + s}.$$

Suppose that we want to find a second-order controller $G_c(s) = (k_2s^2 + k_1s + k_0)/(l_2s^2 + l_1s + 1)$ with design parameters $\gamma = [2\ \ 2\ \ 2\ \ 2.5]$, $\tau = 0.77314$, and a fixed gain $k_0 = 10$.

```
>> ap=[0.25 1.25 1 0]; bp=[1];
>> gr=[2 2 2 2.5]; nc=2; mc=2; t=0.77314; k0=10;
>> [bc,ac,aa,g,tau,gs,rr]=gk2c(ap,bp,nc,mc,gr,t,k0)

bc =
    1.0487   6.7314   10.0000
ac =
    0.0044   0.0922   1.0000
aa =
    0.0011   0.0286   0.3697   2.3910   7.7314   10.00
g =
    1.9998   2.0000   2.0000   2.5000
tau =0.77314

gs =
    0.5000   1.0001   0.9000   0.5000
rr =
    -7.1863 ± 8.2769i
    -3.9071 ± 2.2819i
     -3.6788 + 0.0000i
```

gk2t

Purpose:

Find possible equivalent time constants τ for a given plant (`ap,bp`) by adding a constraint on the steady state gain of controller to `g2t`.

Synopsis and Description:

```
tau=gk2t(ap,bp,nc,mc,gr,k0)
```

Inputs:

`bp,ap`= numerator and denominator polynomials of the plant.
`mc,nc`= numerator and denominator orders of the controller.
`gr`= a vector of the desired stability index.
`k0`= a steady state gain of numerator polynomial of the controller.

Outputs:

`tau`= possible equivalent time constants.

Example:

Consider the following third order plant.

$$G_p(s) = \frac{5}{0.05s^3 + 0.6s^2 + s}.$$

Let us find possible equivalent time constants under the condition of the desired stability index of $\gamma = [2\ 2\ 2.5]$ when a first-order controller $G_c(s) = (k_1 s + k_0)/(l_1 s + 1)$ with a fixed gain $k_0 = 1.0$ is considered.

```
>> ap=[0.05 0.6 1 0]; bp=[1];
>> gr=[2 2 2.5]; nc=1; mc=1; k0=1.0;
>> tau=gk2t(ap,bp,nc,mc,gr,k0)

tau =
    2.9084
    0.5640
   -0.4724
```

gkc

Purpose:

A batch file combining `gk2c` and `c2g`. It provides various design results including CDM controller and their time/frequency responses.

Synopsis and Description:

```
gkc
```

Initial information:

`bp,ap`= numerator and denominator polynomials of the plant $G_p(s)$.
`mc,nc`= numerator and denominator orders of the controller $G_c(s)$.
`gr`= a reference stability index.
`t`= a desired equivalent time constant
`tm`= time scale for time response, which is normally set to 1 or 0.5.
`k0`= a steady state gain of numerator polynomial of the controller.

Outputs:

a plot representing time and frequency responses of $G(s)$, $T(s)$, and $S(s)$.
`ba,bc,ac`= polynomials of the designed CDM controller.
`aa`= characteristic polynomial of the closed loop system
`g`= stability index of `aa`.
`tau`= equivalent time constant of `aa`.
`gs`= stability limit of `aa`.
`rr`= roots of characteristic polynomial `aa`.
`pmgm`= phase and gain margins of the loop transfer function
`wpmgm`= phase and gain crossover frequencies of the loop transfer function

Example:

Consider the same Example as in the function `gk2c`. The outcomes of `gkc` are the same as those in the previous Example and in addition, `gkc` provides the following Fig. A.8.

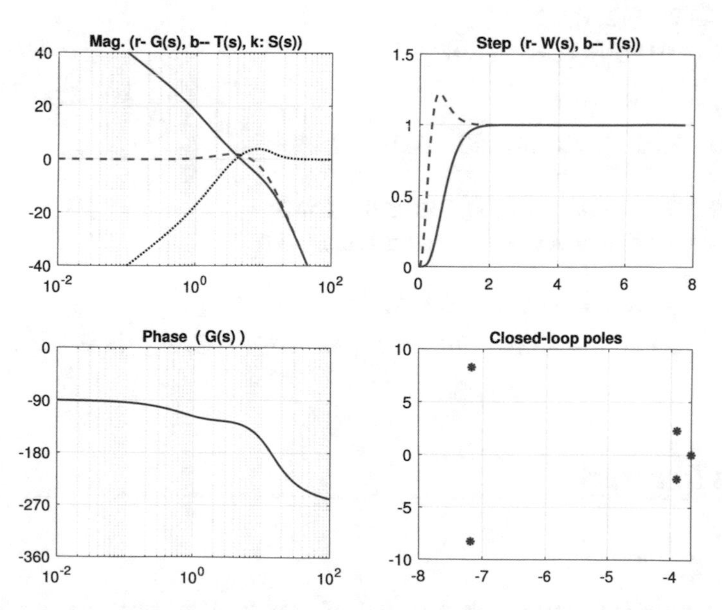

Fig. A.8 Poles, frequency and step responses of resulting systems provided by `gkc`

gt2a

Purpose:

Obtain the characteristic polynomial and the squared polynomial corresponding to the specified stability index and equivalent time constant.

Synopsis and Description:

```
[aa,aq]=gt2a(g,t)
```

Inputs:

g= reference stability index
t= desired equivalent time constant.

Outputs:

aa= characteristic polynomial $P(s)$ corresponding to the specified (g,t).
aq= squared polynomial of aa, which is $A_q = P(-s)P(s)$.

Example:

Find the polynomial and its squared polynomial corresponding to the stability index $\gamma = [2\ \ 2\ \ 2\ \ 2.5]$ and $\tau = 2.5$.

```
>> g=[2  2  2  2.5];  t=2.5;
>> [aa,aq]=g2a(g,an,a0)

aa =
    1.00   8.00   32.00   64.00   64.00   25.60
aq =
    1.00   0   128.00   409.60   819.20   655.36
```

The above coefficient vectors aa and aq represent

$$P(s) = s^5 + 8s^4 + 32s^3 + 64s^2 + 64s + 25.6,$$
$$A_q(x) = x^5 + 128x^3 + 409.6x^2 + 819.2x + 655.36, \quad (x = -s^2).$$

rresp2/rresp3

Purpose:

rresp2 provides a step response plot of two transfer functions specified by $G_1(s) = N_1(s)/D_1(s)$ and $G_2(s) = N_2(s)/D_2(s)$. Similarly, rresp3 is to obtain a step response plot of three transfer functions.

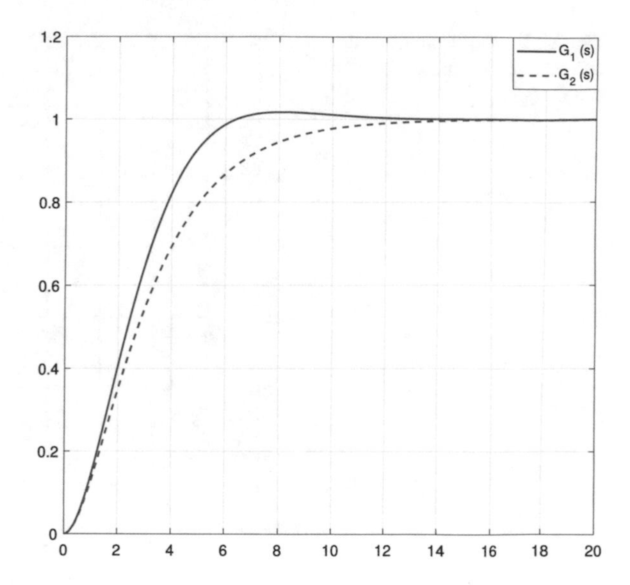

Fig. A.9 Step responses of two transfer functions $G_1(s)$ and $G_2(s)$

Synopsis and Description:

```
rresp2(num1,den1,num2,den2,tmax),
rresp3(num1,den1,num2,den2,num3,den3,tmax)
```

Inputs:

`num1, den1`= numerator and denominator polynomials of $G_1(s)$.
`num2, den2`= numerator and denominator polynomials of $G_2(s)$.
`num3, den3`= numerator and denominator polynomials of $G_3(s)$.
`tmax`= the maximum time to be simulated.

Example:

Obtain step responses of the following two transfer functions using `rresp2`:

```
>> num1=[0.4]; den1=[1 1 0.4];
>> num2=[0.4]; den2=[1 1.3416 0.4]; tmax=20;
>> rresp2(num1,den1,num2,den2,tmax)
```

The output is the step responses plot shown in Fig. A.9.

tresp/tresp4

Purpose:

`tresp` obtains step response of two transfer functions specified by a common denominator $D(s)$ and two numerators $N_1(s)$, $N_2(s)$. Similarly, `tresp4` is to obtain

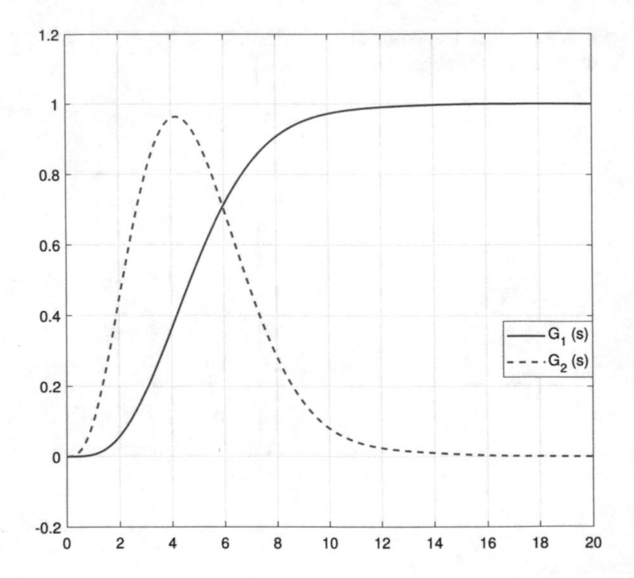

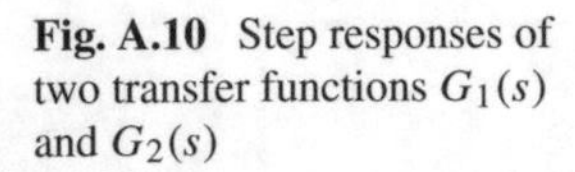
Fig. A.10 Step responses of two transfer functions $G_1(s)$ and $G_2(s)$

a step response plot of four transfer functions specified by four numerators and a common denominator.

Synopsis and Description:

```
tresp(num1,num2,den,tmax),
tresp4(num1,num2,num3,num4,den,tmax, ty1,ty2,ty3,ty4)
```

Inputs:

`num1, num2`= numerator polynomials of $G_1(s)$ and $G_2(s)$ respectively.
`num3, num4`= numerator polynomials of $G_3(s)$ and $G_4(s)$.
`den`= a common denominator polynomials of all transfer functions.
`tmax`= the maximum time to be simulated.
`ty1,···, ty4`= title of each response.

Example:

Obtain step responses of the following two transfer functions using `tresp`:

```
>> num1=[0  0.2];  num2=[1  0];den=[1  2  2  1  0.2];  tmax=20;
>> tresp(num1,num2,den,tmax)
```

The output is the step responses plot shown in Fig. A.10.

References

1. Astom KJ, Hagglund T (2006) Advanced PID control. ISA, Research Triangle Park
2. Ackermann J (1993) Robust control: systems with uncertain physical parameters. Springer, London
3. Brmish BR (1994) New tools for robustness of linear systems. Maxwell Macmillan Canada Inc, Ontario
4. Bhattacharyya SP, Datta A, Keel LH (2009) Linear control theory: structure, robustness, and optimization. CRC Press, Boca Raton
5. Bode HW (1945) Network analysis and feedback amplifier design. Van Nostrand, New York
6. Darbha S (2003) On the synthesis of controllers for continuous time LTI systems that achieve a non-negative impulse response. Automatica 39(1):159–165
7. Djaferis TE (1995) Robust control design: a polynomial approach. Kluwer Academic Publishers, Norwell
8. Fuller AT (1975) Stability of motion. Taylor and Francis, London
9. Hamamci SE, Kaya I, Atherton DP (2001) Smith predictor design by CDM. In: Proceedings of European control conference, Porto, Portugal, September, 2001, pp 2365–2369
10. Hamamci SE, Koksal M, Manabe S (2002) Robust position control of a RADAR antenna with the coefficient diagram method. In: Proceedings of the 4th Asian control conference, Singapore, September 25–27, 2002, pp 1785–1790
11. Hamamci SE, Koksal M, Manabe S (2002) On the control of some nonlinear systems with the coefficient diagram method. In: Proceedings of the 4th Asian control conference, Singapore, September 25–27, 2002, pp 1791–1796
12. Hara S, Hori Y (2000) MATLAB base system for CDM and design example of vibration suppression controller for 2-inertia system. In: Proceedings of the 3rd Asian control conference, Shanghai, China, July 3–7, 2000, pp 2085–2090
13. Hirokawa R, Sato K, Manabe S (2001) Autopilot design for a missile with reaction-jet using coefficient diagram method. In: Proceedings of the AIAA guidance, navigation, and control conference, Montreal, Canada, August, 2001, pp 739–746
14. Hori Y (1996) A review of torsional vibration control methods and a proposal of disturbance observer based new techniques. In: Proceedings of the 13th IFAC world congress, San Francisco, USA, pp 990–995
15. Hunt KJ (1993) Polynomial methods in optimal control and filtering. Peter Peregrinus, London
16. Hwang C, Hwang JH, Hwang LF (2002) Design of PID-deadtime control for time – delay systems by coefficient diagram method. In: Proceedings of the 4th Asian control conference, Singapore, September 25–27, 2002, pp 1178–1182
17. Ikeda H, Kubo K, Yano Y, Inami H, Wakamiya Y (2002) A new tension control method for hot strip mill based on CDM. In: Proceedings of the 4th Asian control conference, Singapore, September 25–27, 2002, pp 1780–1784
18. Isarakorn D, Panaudomsup S, Benjanarasuth T, Ngamwiwit T, Komine N (2002) Application of CDM to PDFF controller for motion control system. In: Proceedings of the 4th Asian Control Conference, Singapore, September 25–27, 2002, pp 1173–1177
19. Johnson MA, Moradi MH (eds) (2005) PID control: new identification and design methods. Springer, London
20. Kim DK, Kim HS (2002) A study of the controller design for pendubot using CDM. In: Proceedings of the 4th Asian control conference, Singapore, September 25–27, 2002, pp 1155–1160
21. Kim YC, Cho TS, Kim HS (2000) Comparative studies of control design method. In: Proceedings of the 3rd Asian control conference, Shanghai, China, July 3–7, 2000, pp 2061–2066
22. Kim YC, Manabe S (2001) Introduction to coefficient diagram method. In: Proceedings of the 1st IFAC symposium on system structure and control, Prague, Czech, August 29–31, 2001, pp 147–152

23. Kim YC, Keel LH, Bhattacharyya SP (2003) PID controller design with time response specifications. In: Proceedings of the American control conference, Denver, Colorado, June 4–6, 2003, pp 2238–2244
24. Kim K, Kim YC, Keel LH, Bhattacharyya SP (2003) Transient response control via characteristic ration assignment. IEEE Trans Autom Control 48(12):5005–5010
25. Kim YC, Jin L (2008) Fixed, low-order controller design with time response specifications using non-convex optimization. ISA Trans 47:429–438
26. Kitamori T (2001) Partial model matching method conformable to physical and engineering activities. In: Proceedings of the 1st IFAC symposium on system structure and control, Prague, Czech, August 29–31, 2001, pp A-127
27. Komine N, Shibata K, Benjanarsuth T, Ngamwiwit J (2002) Weighting matrices selection of derivative state constrained control by CDM. In: Proceedings of the 4th Asian control conference, Singapore, September 25–27, 2002, pp 1774–1779
28. Kumpanya D, Panudomsup S, Benjanarasuth T, Ngamwiwit J, Komine N (2002) PI controller designed by CDM for process with dead time. In: Proceedings of the 4th Asian control conference, Singapore, September 25–27, 2002, pp 1167–1172
29. Golnaraghi F, Kuo BC (2017) Automatic control systems, 10th edn. McGraw-Hill, New York
30. Lai YG, Lin YY, Hung CZ (2002) Control laws comparison via real-time stabilization of two-link inverted pendulum. In: Proceedings of the 4th Asian control conference, Singapore, September 25–27, 2002, pp 1183–1188
31. Manabe S, Tsuchiya T (1984) Controller design of flexible spacecraft attitude control. In: Proceedings of the 9th IFAC world congress, Budapest, Hungary, July 2–6, 1984, vol 17(2), pp 2939–2944
32. Manabe S (1994) Coefficient diagram method as applied to the attitude control of controlled-bias-momentum satellite. In: Proceedings of the 13th IFAC symposium on automatic control in aerospace, Palo Alto, CA, September 12–16, 1994, pp 322–327
33. Manabe S (2001) Diophantine equations in coefficient diagram method. In: Proceedings of the 1st IFAC symposium on system structure and control, Prague, Czech, August 29–31, 2001, pp A-130
34. Manabe S (2002) Application of coefficient diagram method to MIMO design in aerospace. In: Proceedings of the 15th IFAC world congress, Barcelona, Spain, July 21–26, 2002, vol 35(1), pp 43–48
35. Manabe S (2002) A suggestion of fractional-order controller for flexible spacecraft attitude control. Nonlinear Dyn 29:251–268
36. Manabe S (2004) Comparison of H-inf and coefficient diagram method in aerospace. In: Proceedings of the 16th IFAC symposium on automatic control in aerospace, Saint-Petersburg, Russia, June 14–18, pp 394–399
37. Manabe S (2004) Design of fractional order control system under strong influence of saturation. In: Proceedings of the 1st IFAC workshop on fractional differentiation and its application, Bordeaux, France, July 19–21, 2004, pp 676–681
38. Mitsubishi Space Software Home Page (2000) Available at http://www.mss.co.jp/techinfo/cdmcad/cdm_progam.htm. Accessed 10 Aug 2020
39. Pang GKH (2000) Coefficient diagram method toolbox for use with MATLAB. In: Proceedings of the 3rd ASCC, Shanghai, China, July 3–7, 2000, pp 2079–2084
40. Photong P, Kampanaya D, Komine N, Ngamwiwit J (2000) Application of CDM to PIDA control. In: Proceedings of the 3rd ASCC, Shanghai, China, July 3–7, 2000, pp 2073–2078
41. Routh EJ (1905) Dynamics of a system rigid bodies. MacMillan, London
42. Takeda H, Manabe S (1959) The effects of mechanical resonance systems on control loop. J Inst Electr Eng Jpn 79(848):611–618
43. Takeda H, Manabe S, Sakagami S, Hosono I (1961) Control equipment for transonic wind tunnel drives. Mitsubishi Denki Giho (Mitsubishi Electr Tech J) 35(6):1–11
44. Tesfaye T, Ochiai Y (2000) Application of coefficient diagram method to motion control. In: Proceedings of the 3rd ASCC, Shanghai, China, July 3–7, 2000, pp 2067–2072
45. Visioli A (2006) Practical PID control. Springer, London

46. Wolovich WA (1974) Linear multivariable systems. Springer, New York
47. Wolovich WA (1994) Automatic control systems: basic analysis and design. Saunders College Publishing, Orlando

Index

S. Manabe and Y. C. Kim, *Coefficient Diagram Method for Control System Design*, Intelligent Systems, Control and Automation: Science and Engineering 99,
https://doi.org/10.1007/978-981-16-0546-8

Printed in the United States
by Baker & Taylor Publisher Services